OLD ENGLISH GAME BANTAMS

OTHER TITLES

Other titles written by Joseph Batty include the following:

Domesticated Ducks & Geese
Breeds of Poultry & Their Characteristics
Breeds of Ducks & Geese
Artificial Incubation
Natural Incubation
Japanes Long Tailed Fowl
Practical Poultry Keeping
Bantams
Understanding Old English Game
Understanding Modern Game
(with James Bleazard)
Natural Poutry Keeping
Keeping Jungle Fowl

OLD ENGLISH GAME BANTAMS

JOSEPH BATTY

Past President: Old English Game Club

Elected Life Member:Old English Game Bantam Club

Beech Publishing House
The Bindery
Sawmill Buildings
Stedham
Midhurst GU29 ONY

ISBN 1-85736-120-2

Dedication

Dedicated to Game Fanciers in every land, especially to past and present members of the
Old English Game Club (Carlisle) and the
Old English Game Bantam Club

Beech Publishing House
The Bindery
Sawmill Buildings
Stedham
Midhurst GU29 ONY

CONTENTS

PREFACE

This is a concise guide to the extraordinary breed of bantams - Old English Game. They offer great variety in colours and variety and deservedly rank first in the breeds kept by fanciers. They have not been spoilt by commercial considerations and as a result the development of the breed represents many generations of dedication from enthusiastic breeders.

Although the memory of my early days are not vivid there are some incidents which stick in my mind. Starting school, being spoilt by my Grandfather (also called Joseph) and similar events are there in fragmented form. However, I do remember quite vividly being given my first trio of Spangled bantams which were purchased by my father from a farmer on a hill farm in Penistone; I was five at the time. I cannot be sure of their exact worth in terms of exhibition qualities, but they were treasured like the many other forms of livestock which were purchased and kept.

Since that time I have kept Old English Game whenever possible and in many places. This spans a period of well over 50 years and I have kept them *continuously* for the last 35 years. They have given great pleasure and interest; a relaxing hobby with each new year giving further and more varied experiences.

J Batty **December 1994**

ACKNOWLEDGEMENTS

I acknowledge with thanks the assistance given by many people and organizations who have assisted in supplying details or photographs. Wherever possible these have been indicated in the text.

The Old English Game Club gave permission to use the ***Standards.*** However, the explanations and notes are my own and should be viewed as such.

Poultry World and the ***Poultry Club*** have given assistance when approached. Fanciers have supplied photographs and information very willingly and thus have allowed a wide range of opinions to be incorporated into the text. Authors past and present have been consulted and their works have been cited.

The ***American Standard of Perfection*** (American Poultry Association) and ***British Poultry Standards*** (Poultry Club of Great Britain) have been consulted and readers wishing to study further detailed standards are referred to these works.

JB

POINTS OF THE OLD ENGLISH GAME BANTAM
(Prizewinning Black-Red – Trevor E. Thompson)
(Photo Courtesy: *Poultry World*)

O.E.G. BANTAMS

	Left	*Right*
Top	Black Red (Partridge)	Black Red (Wheaten)
Middle	Spangles	Pile
Bottom	Golden Duckwing	Silver Duckwing

Note:
In the illustration the Partridge Hen should be a good even-brown colour. Nowadays Wheaten Hens possess a light hackle.

Charles Francis

1

HISTORICAL BACKGROUND

What is A Bantam?

Traditionally a bantam is a small version of a large fowl or is a naturally small fowl, such as the Pekin. Old English Game Bantams (referred to as OEG bantams from hereon) are not natural bantams, but are the result of selected breeding and crossing with different breeds to produce the types we know today.

In size (mainly weight) the bantam is around 25% the size of the large fowl, but the exact proportion varies from one breed to another and one country to another. Moreover, the weights have varied because skilful breeding has produced sizes which were impossible 50 years ago. No longer is it necessary to breed the chicks very late in the season and feed the minimum of food to reduce growth. The correct size has been realized by careful selection of parent stock. Moreover, it is important to have birds which are fit and plump (but not fat) because they are able to reproduce and breed similar birds, thus perpetuating the strain which has been developed.

The normal size of OEG bantams is in the region of 20oz for females and 25oz for cocks, but these weights are considered further in connection with the *Standards*, which are guide lines issued by poultry clubs and breed clubs throughout the world. Strangely enough, the standard weights have tended to increase for bantams because Entwisle, writing in the 1890s stated that bantams were a fifth the size of the equivalent large fowl. However, in OEG bantams they appear smaller because they are more compact and graceful. Between 1930 and 1982 the recommended weight in the *British Poultry Standards* rose from Cock 18 - 22oz to 22 to 26 oz and the female 16 - 20 to 18 - 22oz. This represented a substantial increase yet the birds are very much bantams, but any further increase does not appear warranted. In the USA (American Standard of Perfection) the recommended weights are 24oz for a cock and 22oz for the hen with cockerels and pullets being 2oz lighter.

There is difficulty in establishing the **normal size** of large OEG because these vary tremendously. Indeed, the large, show type may be as heavy as 6lb or more, but the Oxford type bird may be as light as 4.5lb. Since the concern with bantams is with showing, the higher figure is the logical criterion. A bantam should look elegant, yet powerful for its size.

Exact Date Unknown

The precise date when OEG bantams came into existence cannot be stated with exactness. Nor can we state with accuracy the make–up of the breed. Some suggest there is no large OEG in the ancestors, but we cannot be certain. There is more evidence to suggest large OEG blood than there is of a development from Modern Game. If there was no Game blood in the development why call them Old English Game? The old Game fanciers were dedicated people and I cannot believe that there is no link with the large breeds. Generally those who make such assertions have not been Game followers so they are simply making unsubstantiated statements.

LINKS WITH THE PAST

There are many strands of evidence to show that OEG bantams existed before Modern Game bantams. Moreover, there is general acknowledement that Moderns are show birds developed for that purpose. Poultry, including miniature fowl, have existed for thousands of years. Large OEG are amongst the oldest breed known and therefore it is inconceivable that bantams in Modern form came first; the emergence of shows was a Victorian phenomena and this fact alone gives us the date when Modern Game were created.

The evidence available may be summarized as follows:

1. Authors
2. Poultry Club Standards
3. Comparisons of Earlier Breeds and their link with OEG and Modern Game

All can shed light on the question of the origins. What does seem very clear is that later authors have quoted previous records or books without checking the true facts. The most important books on poultry were written when poultry showing was in its heyday. Lewis Wright in particular, who did such noble work for the Fancy, led later writers to believe that "Game" meant Modern Game. Even his large OEG were Modern Game, painted in an idealized form by Ludlow, so it is not really surprising that later on there were wrong conclusions drawn. In the context of his times he was reporting ***Game*** as they were being shown, although not necessarily as they existed in their entirety, because OEG were still on the farms and small holdings of Cumbria and we have only to look at the works of Herbert Atkinson, the founder of the Oxford Old English Game Club, to see what steps were being taken to bring back the true Old English Game. Indeed, in Cumbria in the 1880s steps were being taken to include OEG bantams in shows (see p 13) and the Old English Game Club was about to be formed (1887). The Oxford Old English Game Club was formed in 1885, but this does not cater for bantams.

Authors

The authors can be divided into three categories (although they do overlap) :

1. **The early years of the Fancy**
2. **The Victorian heyday of the Fancy**
3. **Modern Authors ; ie, 20th century.**

In the first category must come **W Wingfield & C W Johnson**, *The Poultry Book*, published in 1853, who cite the existence of OEG bantams being at the Metropolitan Exhibition in 1852. There was also *The Poultry Book* , by **W B Tegetmeier** (1867), where he shows an illustration of OEG bantams, looking very much like the large OEG, which, despite the rather long bodies, were minatures of the breed. (Fig. 3).

One of the most significant titles is by **Wm. M Lewis** (*The People's Practical Poultry Book*, Moore, NY, 3rd ed. 1871) in which he gives illustrations of both bantams and large OEG. The mania for Modern Game had not arrived in the USA, being recognized in 1874 so the author was not influenced by the show scene in Britain. (These birds are re-produced in Figure 3.4)

Adding to the confusion, in Britain there were advertisements for various Game bantams in *The Poultry Review* (24th January, 1874), a magazine devoted to the poultry fancy. This offered Game bantams of the Brown Red variety.

The journal copy was portraying an OEG bantam-type, but it seemed that the bird being advertised was a Modern-type. References to the "Game" in contemporary show entries makes it clear that these were Modern and not OEG bantams. Study of an advertisement by W F Entwisle, and a cross reference to his book ***Bantams*** (1894/5), makes it clear that this was very much the era of the Moderns. However, and showing the link with OEG, Entwisle gives ***standards,*** compiled by himself, which gives **OEG colours**. These were modified later when Moderns were restricted to a few colours.*

*** The colours given by him were Black-Reds, Brown-Breasted Reds, Golden Duckwings, Red Piles, Whites, Pile Wheatens, and Birchen Greys. Blacks and Blue-Duns now scarce.**

Proceeding in his description Entwisle, who was responsible for creating bantams for many breeds, laments the loss of many of the old colours in Game. He notes that 25 to 30 years previously -- making this around 1860 after allowing for the time for writing his book -- there were Spangles being bred in Southwell, Edingley and Farnsfield, Notts. These were the normal Black-Red Spangles and what he called Grey-spangled. He also states:

> **These colours are frequently seen in Old English Game, and also in Aseels, and we used to have them in Aseel bantams ....**

An interesting description on how he would produce a Game colour makes interesting reading and undoubtedly gives a clue on how he converted OEG bantams to the Modern type. For this purpose he would use an Aseel cock and over a period (4 years is inferred) would select the type and colour until the type was achieved. (see *Bantams, **ibid***).

There is little doubt that Entwisle was providing an overview of the Fancy as it stood in the 1890s (he died in 1892 and his book was published by his daughter), and at that time it seemed the old colours and OEG bantams had disappeared, although this was not the case.

Lewis Wright, in *The Illustrated Book of Poultry*, 1873, did not recognize the existence of OEG bantams and the reason is obvious. Between around 1850 and 1873 the Modern Game breed was evolved in large Game and, although not acknowledged by many writers, the Modern Bantam bantam appeared. Therefore all references to "Game" were to this new type -- hence the name given to them.* From then onwards, until there was a turning back to Old English Game, because they never really disappeared, the Modern was the main bird for showing. In fact, in the 1911 edition of Lewis Wright's book, revised by Sidney Lewer and others, the existence of OEG bantams was finally acknowledged.

Harrison Weir, (*Our Poultry*), writing in 1902, acknowledged the existence of Game bantams. He gave full attention to large OEG, but not to bantams. Around the same time **P Proud** wrote *Bantams as a Hobby* and in that book he admits first hand knowledge of OEG bantams going back pre-1850 to his grandfather's time.

*** Readers are referred to *Understanding Modern Game*, J Batty and James Bleazard.**

Around this time other books appeared which included bantams and Game bantams, some distinguishing between Modern and OEG. **L C Verrey** and **W F Entwisle** were important writers in their day. Bantams had truly arrived and the Game bantam was now divided into Modern and OEG. However, it was not until the latter part of the first decade of the 20th century that the distinction was made clear.

When 1920 arrived the OEG bantam had really established itself as a popular show bird. In *The Feathered World Year Book* for 1920 it was stated :

> **The war does not appear to have had any adverse effect on this breed, for, if anything, it is more popular than ever. The entry that has been seen at almost every show that caters for this sturdy little bird-- and what show does not?-- could hardly have been improved upon in pre-war days. Quality, too, has been well maintained, and some splendid birds have been staged at almost every fixture.**

Now these words do not suggest a new breed and, looking at the practical realities, how could the true OEG bantam have been created sc quickly from the tall, slim Modern Game birds. Entwisle speaks of four years for creating a bantamized breed and my experience suggests this is correct, with at least two more years to eliminate foreign influences so the birds breed true and consistently.

Was it also another miracle that many colours, seen in large OEG were in existence. Many experienced fanciers will confirm that it takes many generations and years to produce new colours. Again quoting from *The Feathered World* report for 1920 we see there were many colours, never seen in Moderns; thus:

> **There is no lack of choice in colour for anyone wishing to take up the breed, for there are Spangles, Black-Reds, Blue-Reds, Creles, Blues, Blacks, Birchens, Brown-Reds, Duckwings, Piles, Furnaces, etc; both plain and muffed to pick from. Spangles are easily the most popular, and although Black-Reds come in a useful second, they are being closely pressed for this honour by some of the off-colours, as they are generally called.**

As noted, when dealing with Entwisle, there were bantams around previously, but did not feature in the conversion to Modern Game bantams. Indeed, Spangles are the most difficult of birds to breed true to colour and it would be impossible to produce these from the limited colours of the Moderns which at that time were limited to Black-Reds, Piles, Duckwings and Birchens, all colours which existed with OEG large and bantams.

Later, modern writers on bantams fell into the same trap of assuming that Moderns pre-dated OEG bantams. Both W H Silk and H Easom Smith took the view that there was no large Game in the make-up of bantams and Moderns came first. It was left to later fanciers to doubt the veracity of statements made on the dating of the existence of the breeds -- Modern and OEG bantams. Fred P Jeffrey in the USA (*Bantam Breeding & Genetics* , second edition, 1977) queries the fallacy that had been created. He notes that books published prior to 1853 contained no references to Game bantams, yet it clear from various sources that they existed. Small fowl, domesticated Red Jungle Fowl, probably came back from India on ships and these would be the OEG bantams kept by fanciers. It is a sound theory.

The author of this book stated in a book

published in 1973, (*Understanding Old English Game*) that OEG were in existence before Moderns. Also in *Understanding Modern Game*, **ibid**, it was stated that there was little doubt that Modern bantams were developed along the same lines as the large Modern Game. Accordingly, OEG were regarded as the older breed.

The Poultry Standards

The ***Standards of Perfection*** were drawn up by the earlier pioneers such as Lewis Wright and W F Entwisle in the UK and others in the USA, and then developed and approved by the breed clubs and finally adopted by the poultry clubs. In the USA the OEG bantam was standardized in 1925. On the other hand, they appeared in Britain in the first *standards* of 1865, but only as smaller birds, which should come as near as possible to the large Game, except for size.

The reader will appreciate that here was a description of large **Old English Game** (not Modern) and Game bantams had to comply with these first ***standards.*** Accordingly, the conclusion must be that OEG bantams were in existence and recognized by those who drafted the *standards*. Modifications came later, after judges and the Fancy created the new (Modern) breed.There is acknowledgement that the evolution came slowly over a period from around 1850 to 1900, when the large Moderns almost disappeared and the bantams continued, as they do to the present day. However, the swing back to keeping OEG bantams was dramatic, due no doubt to the ease of breeding and popularity of the breed.

Comparisons of Earlier Breeds.

The early references to Old English Game bantams indicate that they were in existence before the Modern Bantams. However, what of the birds themselves; how did they develop.?

The evolution of **large** Game is well recorded and shows how the Modern Game came into existence by transforming the normal legged bird of the 1850s into a tall bird with whipped tail in 1870 and then to an extremely large bird with stilt-like legs in 1900, when it reached the peak of perfection.

Bantam development undoubtedly went the same way, but because shows came into existence in the period 1850 onwards, although there was a full record of large OEG, because of cockfighting, there were no similar records for bantams. Eventually, the warning bells sounded and OEG bantams (the true type) were shown in 1883 at Cleator Moor in England. Modern and OEG were now separated.

CONCLUSION

We are now in a position to draw conclusions on the facts presented. These can also be substantiated by illustrations of Game birds, both Modern and Old English, thus tracing the possible origins. The evidence available suggests the following:

1. Bantams or miniature OEG were in existence for many decades before the age of the poultry shows. These were kept by farmers and smallholders and allowed to find much of their own food. This is still seen today; there are many places throughout the country where bantams are kept which resemble OEG.

In passing I must place on record that I have had **large** OEG that were perfect in every way as regards conformation and colour, and were hardy and had strong constitutions. In particular, many years ago a trio of Black-Breasted Black-Reds were obtained from a farmer on Dartmoor. These were magnificent; the male was the darkest red possible and the hens were a Black with a brilliant purple and very dark green sheen. The male bird was very small and subsequent male offspring were always small, around 3.5lb.

To create a bantam from this bird would have been quite simple. What would the likely cross have been? If a natural bantam the choice is limited. Yet, since the bantamized versions were produced from 1850 onwards, the natural bantam appears to be the only possibility. Much can be done to eliminate a cross of a particular breed simply by eliminating those offspring with undesirable characteristics and concentrating on those with the required features. For example, Indian Game (a man-made bantam) crossed with another breed, say, OEG, can within a few years be "bred out" completely, yet leave the offspring with a wider and generally heavier body. We can read the descriptions given by Entwisle (*Bantams* , ibid) and marvel at what can be done in skilful hands. Accordingly, we can look at the natural bantams available and suggest possibilities.

We have Belgians, Booted, Dutch, Frizzles, Japanese, Nankins, Pekins, Rosecombs, Rumpless, and Sebrights (although listed as 'natural' they are said to have been 'created' by Sir John Sebright). This is not to mention the bantams that are found in Japan, excluding those we call 'Japanese', the short-legged variety.

A number of these breeds could have been

used for bantamizing OEG. They are Dutch, Rosecombs, Nankins and Sebrights and, no doubt, even by just careful selection the feat would have been possible. The soft feathered bantams and those with unusual features would have been best avoided. The hard feathered breeds mentioned would have been quite suitable. For example, if the Sebright had been taken, which might have been created from a Henny Game cross any way, the reduction in size would be immediate and then the rose comb would be eliminated within a few generations. Whether this was the basis for the creation of OEG bantams - Sebright X Old English Game - we cannot be certain, but this would certainly be a shorter way than by taking Modern Game bantams with their small bodies and long legs.

2. The writers of the day used the description "Game" to cover both Old English and Modern and because Modern-type bantams were being developed all the concentration was on this new breed, which was a variation of the true Old English.

In the pages which follow some idea of the development can be seen.

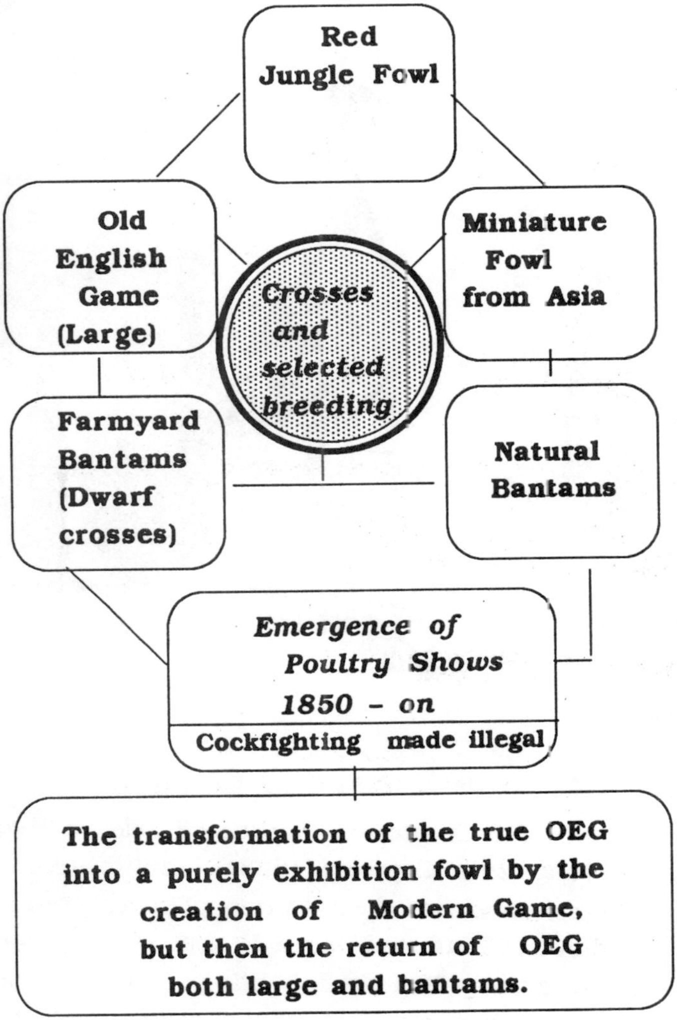

Figure 1.1. Development of OEG Bantams and the Exhibition of Poultry

Figure 1.2 "Modern" Pile OEG Bantams

Drawn by F J S Chatterton in 1888 (*Fowls for Pleasure & Profit*, 25th October, 1888). Chatterton drew what he saw and not the idealized pictures produced by Ludlow. Chatterton stated that these Game should be exactly like their fighting, bigger brethren in everything but size. It will be noted that these specimens are not far removed from the OEG bantams and far from the tall, slender, small bodied birds of later years which became the Modern Game bantams.

Figure 1.3 Early Old English Game bantams
Around 1860 from The Poultry Book, by
W B Tegetmeier, 1867

Figure 1.4 The present-day Modern Game

These bear only a slight resemblance to the earlier Modern bantams (Fig. 1.2) which have continued to develop to the present day.

Figure 1.5 OEG Black-Red Bantam cock
Winner at Birmingham & Carlisle © 1904

Figure 1.6 OEG Bantam Partridge Hen

Owned by Rbt. Pashley. Winner at Crystal Palace & Special Trophies 1906/7/8 and others.

Figure 1.7 Evolution of Modern Game

This depicted large Game, but was the same for bantams

Figure 1.8 Spangle Bantams
Possibly painted by Herbert Atkinson

INTRODUCTION TO OEG BANTAMS

HARDY BIRDS

Old English Game bantams are very hardy and fit. Accordingly, they need no really special management, but the fancier must understand the type of bantam he is dealing with and base his management on what is best for the birds.

First of all there must be appreciation that OEG are hard-feathered; ie, have feathers which are without excessive fluff and are close fitting to the body. This means that in feeding, when getting into condition, the birds must be fed hard corn. The old time Game fancier fed nothing else, but I believe this can be a mistake at breeding time and when birds need vitamins and extra protein to sustain them in the winter or to bring them into condition in the Spring.

Active Birds

Because OEG are active birds they should be allowed ample room for exercise and if kept in a shed or run, especially in the Winter months, they should be given leaves, peat moss and other suitable litter in which they can scratch. If the litter is edible so much the better, because this interests the bantams. If a barrow-load of weeds and other matter, such as soil, grass clippings, and moss are put into a pen just watch how they scratch around, picking out any tasty morsels and gobbling them up. Any birds which do not react in this way are unlikely to be of great value. They are sluggish, over-fed or out of sorts and will never breed successfully, or win at shows. **Constant activity is part of the natural behaviour and should be present.**

Bold & Friendly

OEG bantams are friendly, likeable bantams, but do not be surprised if an over-keen cock "has a go" at other males or even a person who seems to be trying to take his hens. This cheeky aggressiveness is part of the natural make-up and should not be discouraged. The cock should also be a very polite suitor; cocks which grab the food from their wives are not in this category and those which knock them out of the way in a scramble for food less so.

Reasonable Layers

OEG bantams vary in their capacity to lay. I have found some strains lay quite well for a bantam; Partridge/ Black-reds and Piles are generally quite good. However, do not expect commercial egg production and the laying period may be limited to the Spring and Summer. They are bred for shape and other show requirements, but if artificial lighting (25w bulbs) is given to augment light available for a total of 12 hours a day and layers' pellets are fed, a very commendable performance would be possible, in excess of 100 eggs per year. However, bantam-keeping is a hobby and fanciers would be horrified at such practices. It would be rather like fastening your pet dog to a tread-mill all day and then expect him to be a 'friend' in the evenings. The exact number of eggs is immaterial, provided a reasonable number are produced.

The eggs can be utilized for cooking or eating in the usual way, but have two instead of one normal-sized hen egg. Often they are free range so all the better. Remember though, all the eggs from the selected breeding stock should be saved in a cool (not freezing) temperature and incubated whilst not more than a week old.

A decision on whether to use the hen or an

incubator, will also have to be decided. Much depends on the size of the operation and whether you want to hatch a considerable number of chicks. Bantams make good mothers, but the number each can hatch is small, and whilst she is sitting she is not laying. On the other hand, a spell of sitting does reduce the fat on a bantam and to that extent is therapeutic. A Game fancier I knew years ago always insisted that a spell of broodiness reduces the fat in the body and led to a healthier bird. There is no doubt that many OEG bantams are far to heavy and after a few years become short of breath and unhealthy. The aim should be to have a plump bird, but never over-weight.

Standards of Perfection

All breeds of poultry which are "pure breeds" or "standard bred" must comply with the authorized description laid down by the poultry club of the country in question or a specialist breed club which caters for the particular breed. Since the regulations vary from one club to another a reader wishing information should enquire from the poultry club in his or her own country. Generally there must be proof of a new breed being in existence for a number of years and the fact that this is distinct and breeds to type on a consistent basis. Moreover, any new breed must be quite distinct in a number of respects from existing breeds. Differences in colour or some characteristic which is not a fundamental difference makes up to a variety and not a separate breed. Thus in OEG bantams we have:

1. Many different colours*.

These vary from the Black-Red, the colour of the Red Jungle Fowl from White right through to Black. There are even different types of Black, some Blue-bred and others, which show purple and green sheen, which come from a different combination of genes. In all there are around 30 different colours, but only about half that number are seen regularly and are recognizable even by knowledgeable Game fanciers. This aspect is covered in more detail in the chapter on *Standards*.

2. Accepted, natural features found in large OEG.

These are *Tassels*, an appendage of feathers on the back of the head, looking like a tassel, and *Muffs* or whiskers well known in the days of cockfighting (large OEG) by a variety known as Bailey Muffs, bred by Dick Bailey of Carlisle. *Hennies* (Cocks are hen feathered) are also bred; I knew a farmer in Plymstock who bred all the colours in bantam Hennies

*** It should be appreciated that not all colours are recognized in the *Standards*. The American Poultry Association recognizes 19 colours and the Old English Game Club lists 14 and British Poultry Club gives 9, but the off-colours make up the differences.**

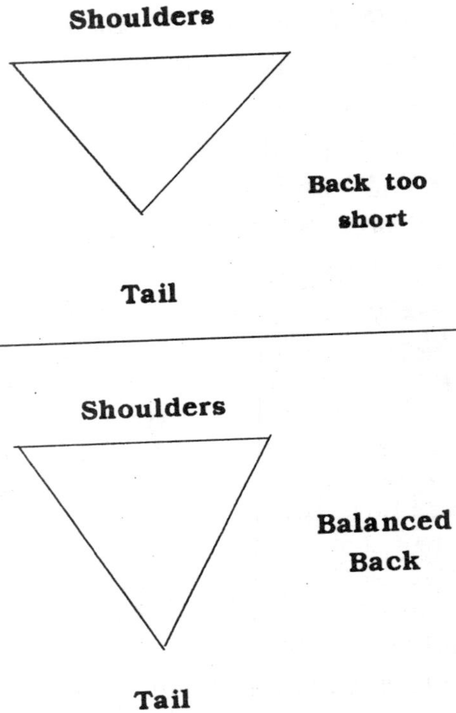

Figure 2.1 The triangular shape of the OEG bantam.

In reality the back, viewed from the top should be heart-shaped, which in essence is a rounded triangle. If too short the bird lacks balance and other essential features would be out of character. For example. if the back is extremely short the wings would tend to look too long for the body. Hens can also walk with a shuffling motion because the legs are 'ill fitting'.

Poultry breeders have quite rightly stressed that the feather type and formation constitute the standard, but this is only a half-truth, yet the feathering is fundamental to the description of Game bantams. If a Game bantam is loose in feather or has too many feathers and 'fluffy' he fails on being a hard feathered bird. Accordingly, any OEG bantams which exhibit too much feather should not be bred from or given prizes. The lighter coloured birds tend to be softer in the feather so we have Whites and Piles which can fail on this score, and even Blues and Blue-bred Blacks can be rather soft in the plumage. In assessing the correct condition, it is necessary to consider the time of year and whether the birds are breeding.

Other factors which must be examined in relation to the *standard* are the shape of the body, eye colour, legs, head and neck, tail, carriage and overall conformation. A sprightly, zestful bantam is called for, not a bird which rests all day on a perch waiting for food. OEG bantams have a style all of their own and if this is not present the breeder is wasting his time.

We are looking to see a well-balanced bird with a compact body, wide at the shoulders and very narrow at the tail.

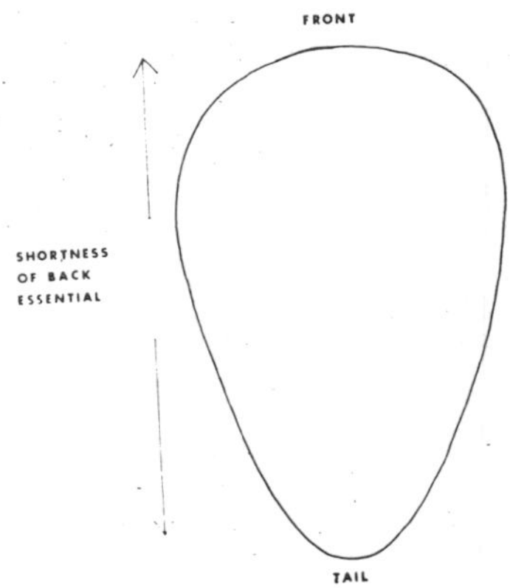

Rounded Triangle to show the Back

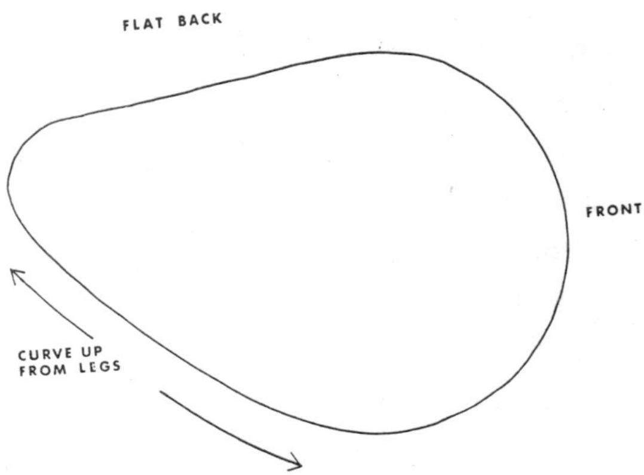

Essential Cone Shape (viewed from side)

Figure 2.2 Basic Shapes of OEG Bantams

The back should be appreciably short, so that between **the base of the neck hackle** (the shawl*) and **the base of the tail** should be relatively close together, but should ***not*** be too close or the bird resembles a fantail pigeon which is quite wrong. If the width of the shoulders and the angle of the triangle formed by the body are about equal this is the ideal. When too short the body is out of balance and the bird appears rather ungainly; like all exaggerated points 'shortness of back' is often given too much prominence, yet a long back is quite uncharacteristic.

If the analysis of the correct shape is taken further we progress to the "cone shape" and the "rounded triangle". For the latter we must imagine the curved shape that fits into an equilateral triangle for this gives us the guide to the ideal shape. Reference to Fig. 2.2 will show what is meant although the example given could be a shorter depth to emphasize the principle. The cone shape, also in Fig. 2.2., is the same shape viewed from the side.

*** Note that a hackle which is too feathery and loose, extending too far on to the shoulders and across the breast is also known as a "shawl hackled" and is a serious fault.**

If taken too far the shape becomes cylindrical or ball-shaped and then, it is argued, the bird is out of balance and does not conform to the ideal shape when examined as a Game bird, with its origins deriving from fighting.

Getting a bird with a flat back, combined with the heart shape across the top can be quite difficult to achieve. Sometimes the ideal is obtained, only to find that the full breast is lost. Also the curve up from the legs, free from fluff and excess feathering, may be very difficult to attain. A bird with a relatively broad body can suffer from a rather protruding belly, and whilst this may be regarded as natural for a laying hen, its absence is more satisfactory and can in reality be achieved. I have bred many varieties without this problem, but equally I have possessed and seen hundreds of bantams which have an excessive bulge under the legs. Sometimes this is due to bantams being too fat, but often it is a characteristic fault in the strain and birds showing this fault are best left out of the breeding pen.

Opposite

Figure 2.3 Tassel & Muff OEG Bantams

Top: Muffs: Silver Duckwing cockerel (J Woodward)
Bottom: Tassels: Detail from a painting by Herbert Atkinson

Choice of Colours

With the many colours available the new-comer to the Fancy must decide which colour he would like to start with. Obviously, this must depend to some extent upon birds being available, but a visit to a local show will usually show which colours are in the area and can be obtained, often birds which are in the Selling Classes. For the novice it might be better to consider a popular colour first and get experience with breeding, say, Black-Reds.

Above all make sure the birds are sound in health and they conform to the *standard*. If a good start is to be made do not buy stock which is quite inferior. Breeders will not usually sell the best stock, but if from sound birds the birds purchased may breed improved bantams. On this basis a new strain can be developed.

3

DEVELOPMENT OF THE OEG BANTAM

Exact Beginnings Obscure

We acknowledge earlier in the book that the exact time of development and their precise origins are obscure, but it is nevertheless believed by the author of this book that OEG bantams are miniatures of the large OEG which have been modified by selected breeding into the types of birds we find today. Undoubtedly, there has been careful breeding to **type** and there has been selected crossing with other breeds to get desirable features; for example, as an experiment, I crossed OEG with Indian Game and as a result got a much broader bird. However, this suffered from overhanging brows (beetle brows), short, stout legs, traces of lacing and a yellow eye. All of these characteristics were undesirable and therefore had to be eliminated without losing the gain in body width. This was achieved in three seasons and the birds were much improved. Obviously the Indian Game were hard feathered birds anyway and were bred originally from a cross which included Old English Game so they were related.

Possible Faults in OEG bred from Indian Game

1. Beetle Brows
2. Eye Colour (Pale: yellow/light)

3. Too much fluff
4. Wrong angle to legs

Figure 3.1 Indian Game Male

width of body of many OEG was 'improved' by use of Indian Game

The Bantam Or Dwarf Gene*

Researchers have shown that there is a gene in many large fowl that will produce dwarf specimens and these can form the basis of bantamized stock. The exact nature of the gene is not well understood, but it is present in some breeds. **There is no difficulty in breeding large fowl with small poultry and thereby developing a bantam.** Undoubtedly, this process has occurred quite automatically in the farm yards of yester-year with the result that miniature OEG have been on farms for many generations.

Starting from OEG bantams, produced as discussed, there was the movement towards more reach, longer legs and so on, until the Modern Game was produced; the whole process was, in fact, **contrary to the requirements of the** *standards* and inevitably these had to be developed for Moderns.

After the turning back to OEG bantams, the process was accelerated from fanciers and out from the farmyards and smallholdings came

---

*** There are many references to corroborate the existence of dwarfing genes. See Hutt F B, *Genetics of the Fowl*, McGraw Hill, and Jeffrey Fred P, *Bantam Breeding & Genetics*, ABA.**

the bantamized version of OEG. The transformation was unbelievable; one moment the "breed did not exist" the next they was a reality and **to a very high quality**. ***This could only come from existing stock.*** It took 50 years for Moderns to reach top perfection so OEG bantams could not be produced from Moderns in a relatively short time.

The son (J F) of F W Entwisle the pioneer bantam breeder has stated that OEG bantams were created from the 'left-overs' from the breeding of Modern Game. He claimed that the colours were Duckwings, Black-Reds and Brown Reds, together with a cross from an Aseel Spangle*. In my opinion this proves nothing; where did the Moderns come from in the first place - my view and that of others is: **from OEG either large or small specimens, no less****.

Since OEG bantams came on the show scene at the turn of the century, with the numerous colours given earlier, it seems surprising that they are regarded as the original Game bantams. There is nothing to substantiate the claim. At the Kendal Show, reported in *The Feathered World Year Book* for 1920, the first

*** Quoted by H Easom Smith in *Modern Poultry Development*, Spur Publications, now op. ** Bewick's drawing of poultry © 1900 clearly shows bantamized poultry amongst large as Farmyard Poultry. See *Old & Rare Breeds of Poultry* J Butler, where the drawing is reproduced.**

Figure 3.2 An OEG Duckwing Bantam

Winner of many prizes including First at Birmingham in 1908 (Mr J R Crompton)

the first issue after World War I, there were 24 classes with a total of 367 entries. The colours were Spangles (93 entries), Black-reds (81), Blacks (37), Brown-Reds (41), Duckwings (16), Self-Blues (36), AOC (35) and Selling (28). At this very early stage the names that stand out are the partnership of J G and T M Hartley a name that was to continue in OEG for a very long period. They won at Kendal, Manchester and Bradford. I remember meeting Tom Hartley at a Royal Dairy Show when I was a young man. How time flies!

From that time on the breed very quickly left behind the Modern Game in terms of followers and birds. There had been a swing back in both large and bantams to the traditional type of bird.

The Early 20th Century Birds

The early OEG bantams showed considerable style. All the ingredients were present from early in the century. If reference is made to Figs 1.5 and 1.6 depicting a Black-Red cock and a Partridge hen respectively it will be seen that these birds were very well developed and undoubtedly were true miniatures of large OEG.

If the analysis is taken a step further a comparison can be made with a large OEG cock and the similarity is very clear evidence that the

Figure 3. Spangle Bantams in 1931
A painting by Herbert Atkinson:
look at the resemblance to the large OEG

bantam contains the blood of the large Game. The conclusive piece of evidence comes from an American source where, in **1871,** both large and bantam OEG were portrayed and are shown below with the original captions.

BLACK-BREASTED RED GAME BANTAMS.

PAIR OF BLACK-RED GAMES.

Figure 3.4 Bantams & Large Game in the USA in 1871 (Source: ***The People's Practical Poultry Book***, **Wm M Lewis, 3rd ed. D I T Moore, NY)**

Figure 3.5 Detail from a drawing by Ludlow.
This was at the turn of the century. There is no resemblance to OEG in the Modern Bantams shown

CONCLUSION

There is little doubt that the OEG bantam existed long before the Modern Game bantam. All the evidence points to this conclusion.

The confusion arose from the lack of written evidence, linked with an upsurge of interest in Shows when Modern Game, large and bantams, were in vogue. As a result OEG bantams were forgotten.

The ***standards*** related to OEG quite clearly, including bantams, and it was only later that this fact was accepted. However, in the book by Lewis (p44) the English standards are quoted and there is no mistaking the fact that the bantams were OEG and not Modern.

Even today there is considerable confusion with bantam ***standards*** departing substantially on some features. Moreover, the British judges follow one set of 'rules' and those of the USA and Australia follow others when dealing with the tail and other characteristics. The time has come when standardization is desirable and the fact that bantams are not fighting Game should be recognized by modification of the ***standards***.

The next two chapters deal with the ***standards*** as they exist. The reader is advised to study these, together with the explanatory notes.

4

STANDARDS I : DESCRIPTION

BASIS OF THE STANDARDS

In Britain the *standards* for bantams have been derived from those laid down for large Game. **The Old English Game Club** with its origins in Cumbria lays down its own version; **The Oxford Old English Game Club** has its *standards* laid down on the basis of the fighting Game bird of olden times.

The **Old English Game Bantam Club** follows a standard which is based on the Oxford ideal.

In the USA the *standards* are somewhat similar to those specified in Britain which is hardly surprising for a breed which is British through and through. This is also the case in Australia.

There are, however, differences on various features, especially the tail and the compactness of the body. In the USA and Australia the *standard* followed is rather similar to that illustrated by Herbert Atkinson around 1930, in

Spangles and Duckwings.

Essentially all agree that the bantam should comply as near as possible with the large ideal, but also state that in reality this is not achieved and therefore allowances have to be made.

The descriptions which follow are based on the Old English Game Club (Carlisle) *Standards*, which are annotated where appropriate. The *standards* are reproduced with kind permission, but the comments are my own based on experience and the views of leading fanciers.

Head: Strong, bold, medium length*.

Note: The head must appear strong in relation to the size of the bird. On no account must he or she appear delicate as shown by a small head on a slender neck. This is the requirement for Modern Game – the head must be long, narrow and slim.

Eyes: Large, bright and prominent, full of expression and alike in colour.

The eyes should indicate alertness and in the Oxford *standard* 'fiery' is included. However, the *standard* does not go far enough in explaining that

***The first part is the *standard* and the following comment or 'Note' is my own: Author.**

the eye should be a specific colour; red in most of the varieties of bantams, but dark for the darker colours such as Brown-Reds. The existence of a very light eye, either pearl or yellow, is a certain sign of Aseel or Indian Game blood and should disqualify. Another undesirable feature, also indicating Asian blood, is the overhanging brow, known as 'beetle brow'. If this condition exists then it is likely that the head is deep and thickset following the manner of Indian Game.

OEG Bantam
Strong without coarseness.

Indian Game
Thick and strong with beetle brow.

Modern Game Bantam
Narrow & snakey

Figure 4.1 Comparison of Heads

Comb: Small, both in cock and hen, and serrated at its edge, single, erect and of fine texture.

One of the problems in judging OEG is that the compulsory dubbing (cutting off the comb and wattles) may hide what had been a twisted or beefy comb, even one which is not evenly serrated. If possible, such birds, having serious faults in the comb, should not be allowed in the breeding pen because the problems will exhibit themselves in the offspring and in the females will certainly show.

Wattles: Fine texture and small.

Ear-lobes: To match the comb and wattles as nearly as possible.

Notes: The ear lobes and wattles should also be of fine texture and be red, the former not exhibiting any white colour. The white often appears in cocks and, again, the dubber's scissors may eliminate the fault.

The dubbing of the comb is essential for all male OEG and it should not be thought that this is a cruel process for it has always been done and, in the many hundreds of birds I have dubbed only once have any ill effects been shown. *The dubbing is described in the chapter on showing.*

Neck: Long and very strong at junction with the body.

All descriptions are relative, but today the bantam does ***not*** have a long neck. This would look out of place on the very compact body of the OEG bantam. A 'medium' neck would be more appropriate, especially also when we look at the now quite small tail, minus large sickle feathers; with a large tail the whole shape lacks harmony. The large Game, with their strong tails look natural with the long tail, but not so on the modern-day OEG bantam in Britain. In the USA and Australia the stronger neck does not come amiss, although the tails are now becoming smaller.

The neck should curve gracefully and not as upright as large OEG; remember, the latter should still resemble the warrior of long ago, upright, proud and defiant; "full of a raging pride" as so aptly described by the old Game men.

Hackle feathers should be hard and fine. The Oxford *standard* states that these should be wiry and cover the shoulders. OEG bantams have short hackles which follow the curve of the neck. As to 'wiry', if the feathers are too scant, late in the season they show wear and tear. Many bantams are not fit for showing at the summer shows because the hackle is distinctly out of condition. It is advisable to spray the birds at regular intervals or mite will spoil the hackle on a cock which tends to be wiry.

Breast, Back and Body: Short back, broad across the shoulders, tapering well to the tail, with a full, broad, well rounded chest, showing as little keel as possible. The keel or breast bone to be straight and of medium length, tapering well up behind, giving a very small and compact body. The whole body with wings as seen from the top to appear flat and near heater shape as possible.

The body shape is of vital importance and readers are advised to study the diagrams given earlier on the essential shape. The reference to a 'heater' may be intended as a description of a flat iron using for pressing clothes which is heart shaped for certainly this is the correct interpretation.

If the wings are strong and positioned so that they are high on the body this gives a more positive look to the shape, Cocks which have wings which hang low at the shoulders never get the desired effect.

The **breast or chest should be very prominent,** but not like the Fantail pigeon whose head rests on the cushion of the tail and cannot even see when he is strutting! This would be quite wrong for OEG bantams. The carriage should be ***fairly*** upright because this emphasizes the prominence of the breast. Long birds tend to be too horizontal and this is a defect.

Figure 4.2 Some Essential Shapes
Bird displaying prominent Breast

The keel should be absolutely straight. Run a finger and thumb along the length of the breast bone and make sure there are no indentations - usually caused by allowing young birds to perch too early or on unsuitable (wide or square) perches. More seriously, it may be a sign of unsuitable upbringing, such as lack of proteins or vitamins in the foodstuff or from being kept indoors without adequate exercise. If the keel juts out below the curve of the breast this is a sign that the bird is not complying with the *standard.*

The back should be as flat as possible without any hump at the point where the thigh bones attach to the body **(roach backed)**, nor should it be narrow and sloping, giving the impression of a weak constitution

Figure 4.3 A Pile Bantam Cock

Good, except keel prominent, spoiling excellent shape.

Wings: Full and round, inclining to meet under the tail, with strong prominent butts, feathers to be fairly broad and furnished with hard strong quills. The primaries not to be too long and to be nicely rounded at the ends and to project past the body as little as possible.

The butts give the shoulders an appearance of great strength and emphasize the width of shoulder. It is also important that the butts are fairly high and not 'off the shoulder' for this gives the impression that the whole wing is too low. In such birds, when handled they appear excellent, but once put down the wing assumes its natural position and spoils the overall compactness.

Strong feathers are called for and none should be missing. In shows broken or damaged feathers should call for penalties. Some weeks before a major show it is wise to examine a bird to ensure that feathers are in sound condition and any broken feathers should be removed to allow for new replacements to grow. However, try to keep the removal down to a minimum. Growing new feathers takes energy and out of the moulting season they may not grow back very evenly.

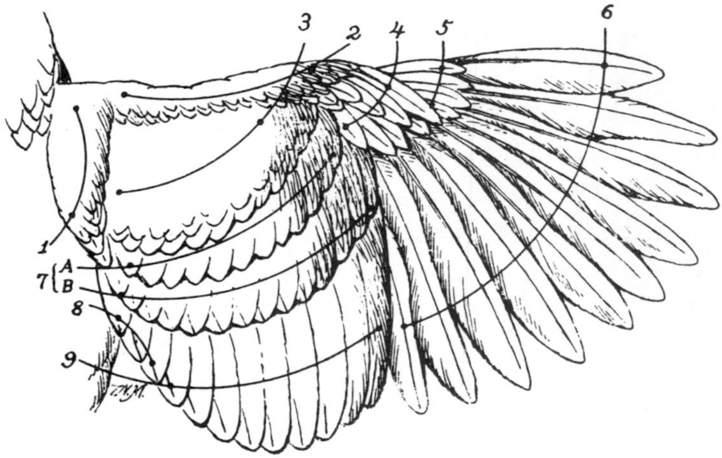

1. Shoulder
2. Wing Front
3. Wing Bow
4/5. Wing Coverts
6. Primaries (Flights)
7. A/B Wing Coverts (Bars)
8. Tertiaries
9. Secondaries (wing Bay)

Figure 4.4 Main Features of A Wing

This is a *general description* of the wing of a fowl (not OEG)

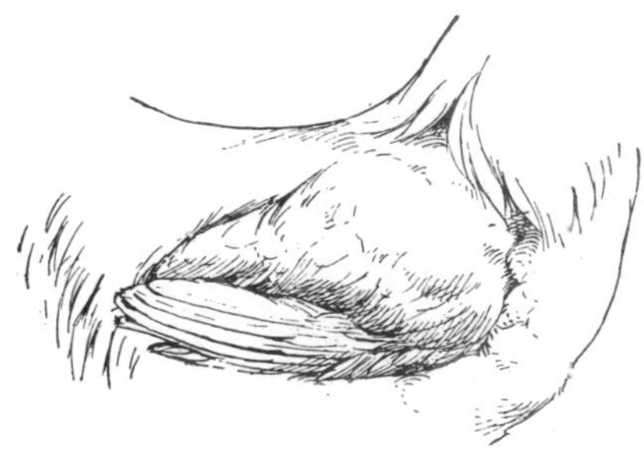

Figure 4.5 *Top* : Wing of OEG Bantam in Moult. (May be mistaken for faulty wing [uneven feathers], but due to moult. Split wing would have larger gap between primaries and secondaries (see Fig 4.4)

Bottom: Loose Wing.

A bird with feathers which extend beyond the body always look out of balance, yet if the wings are too short there is not the streamlined, curved finish to the rear of the bantam.

Getting this curved finish can be difficult and sometimes the very short-backed birds do not achieve the desired result. A bird with a little extra length, giving a narrow rear, can look much more effective than the bird which does not have the compact and well placed wings.

Split wing is an abomination and birds which have the feather formation divided in the middle as a permanent feature should not be given prizes or bred from. When the wing is opened there is a wide gap in the structure of the feathers and, as a result, the wing tends to hang loose, not springing back. Related to this condition is a "loose wing" when the wing feathers tend to show below what should be a neat line where the curve of the wing comes under the tail.

In some birds the wings may reach the thighs and this should be regarded as a fault in the British bird, but is acceptable in the USA and Australian bantam where larger and fuller wings are in order. This was the case in the UK in the 1930s, as shown by the illustration of Spangles by Herbert Atkinson, but, nowadays, the British bantam is no longer expected to have wings to protect the thighs as in the days of cockfighting. Instead they should be positioned to allow a reasonable space to show the thigh, but not high on the body, known as 'Goose-winged'.

Tail: In the cock to be carried at a nice angle, neither too low or too high, and to be straight. Wry and squirrel tails to be considered a serious defect. Feathers to be broad and strong, with a pair of good curved sickles of fair length and well furnished with side hangers. In the hen, well carried, fairly close, and of medium length.

The tail, its size and the way it should be carried present the most controversial area of the *standards*. At the outset, it has to be understood that the tail of the bantam does not display the same height or angle as the large OEG, even in the USA and Australia - where a larger tail is expected. Breeding for a smaller and narrower tail has resulted, in the UK, in the tail being fairly insignificant and the tail lacking in sickle feathers and side hangers, yet for the bird that is now shown, this looks natural and it would appear to be foolish to try and turn back the clock. With a larger tail there would also be a stronger base for the root of the tail and the rest of the body would have to be toned up to match. Study of large OEG and a comparison with the British type of bantam will indicate what is meant.

The USA *standard* specifies that the tail should be set at a 45° angle which avoids the tail which is too high or too low. However, they are dealing with a much fuller tail and this cannot be interpreted too precisely when dealing with the British type.

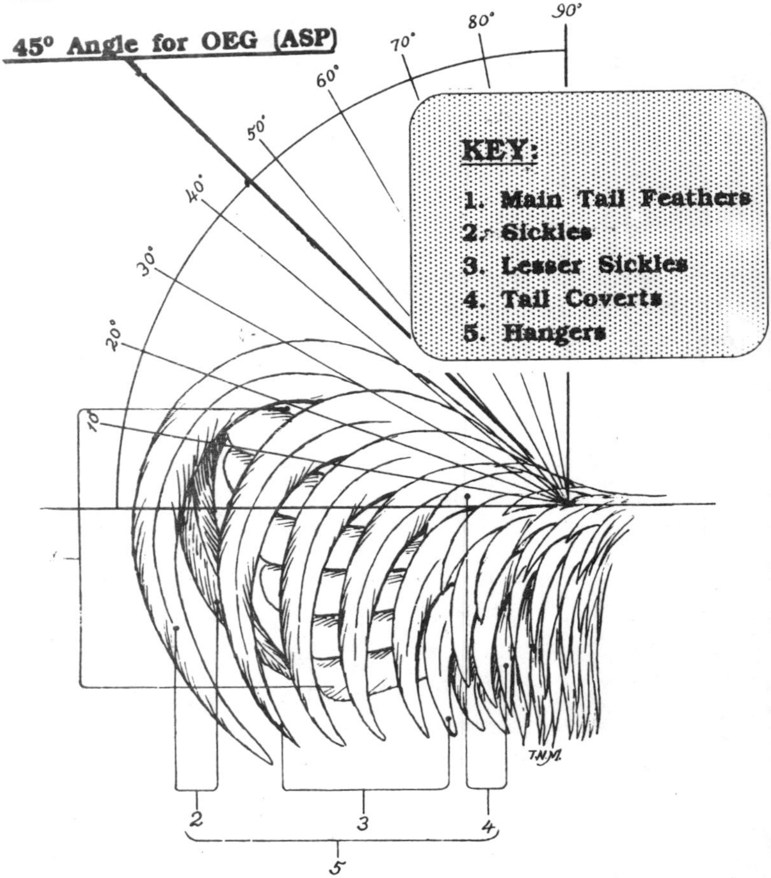

Figure 4.6 Tail of Cock showing Angle
This is intended to show the tail in a general fashion to illustrate the principle involved. It is not an OEG tail. (see Fig. 4.7)

Figure 4.7 Black & White Splashed Male
Note the angle of the tail; many specimens today do not have such long tails.

Legs: Thighs short, thick and muscular, and set fairly well apart; shanks of medium length, with good, round bone; not flat on shins. In-kneed or bow-legged to be considered a serious fault.

This is another area of difficulty. A reasonably tall bird generally presents himself better than a short, squat bantam, especially if there is a bend at the hock, giving the appearance of readiness. The reference to 'set fairly well apart ' should not mean that the legs are stork-like, neither should they be cow hocked. The USA standard states well apart and straight when viewed from the front, but this would make them 'stork legged' and not in accordance with the OEG ideal, which is a bend at the hock following the line of the body, with the spurs almost touching when fully grown - a condition known as 'close heeled', much quoted by the early Game enthusiasts.

In the adult male, over 6 months, there should be spurs which are a reasonable length and fairly fine. (see Below).

Scaly legs and other deformities are serious faults. Regular attention to washing the legs is vital, and any sign of scaly legs should be treated with promptness because once the scaly leg mite have got established the legs take a long time to recover. An ointment which combines sulphur and tar, obtainable from the chemist, quickly removes the parasite.

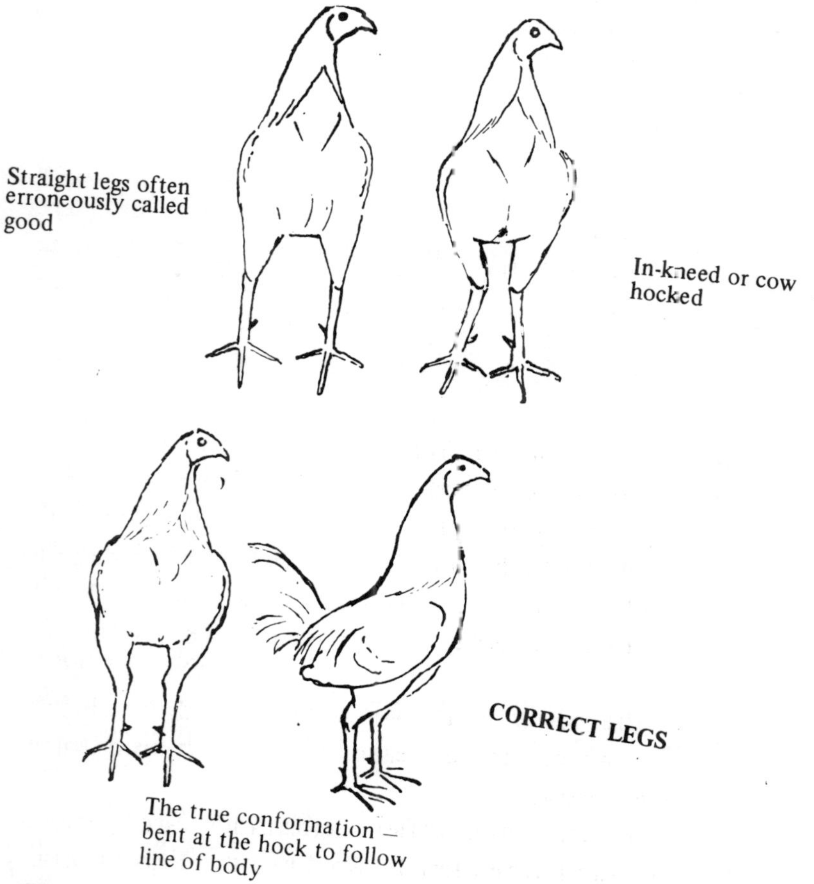

Figure 4.8 Essential Leg Stance (Herbert Atkinson)

The length of leg would be different in the bantam, but the principle is the same.

Feet: Four toes on each foot; should be clean, even, long and spreading, the back toe standing well backward and flat on the ground.

The feet should appear strong and be free from twisted or otherwise faulty shapes, such as bumps at the joints. Older birds, over 3 years of age, may get swollen joints and whilst this may not interfere with their breeding abilities, it does penalise them for showing. The feet and legs set off a bird and therefore should receive regular attention. Even if the bantams taken to a show are not washed fully, and it may be possible to avoid this with dark colours, it is still vital to wash the legs and feet. Nails should also be scrutinized; a bird which is scratching all day will keep the nails short, but a bird in the training pen may have nails which are too long and should be cut with strong nail scissors, but not too short as to cause bleeding.

Duck-footedness is one of the more serious faults in Old English Game. This occurs when the back toe is set close to the foot, thus failing to give the necessary support and impetus to a fighting bird. It is therefore a legacy of the requirement from the days of combat, which, although now illegal, still count when viewing the ideal bird.

Sometimes a bird has a back toe which does not touch the ground when the bird is standing. This is a condition often seen in Modern Game, when the bird stands on his tip-toes. It is a fault in OEG and should be penalized.

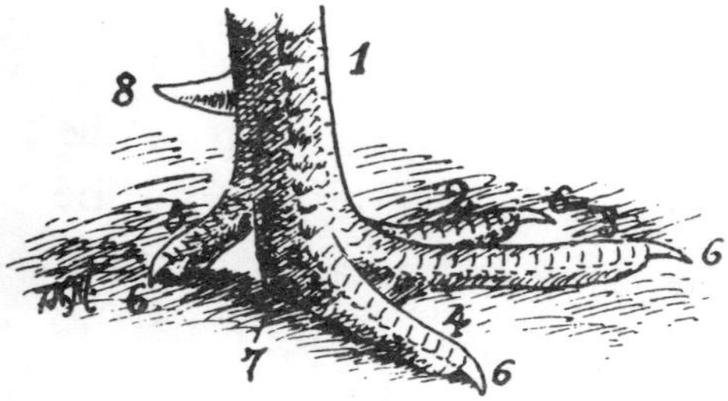

1. **Shank**
2. **Inner Toe**
3. **Middle Toe**
4. **Outer Toe**
5. **Hind Toe**
6. **Toe Nail**
7. **Foot**
8. **Spur**

Normal Foot of Cock

OEG should have the hind toe at an angle from the foot thus giving full support to bird. When close to inner toe this is known as Duck Footed and bird should be disqualified.

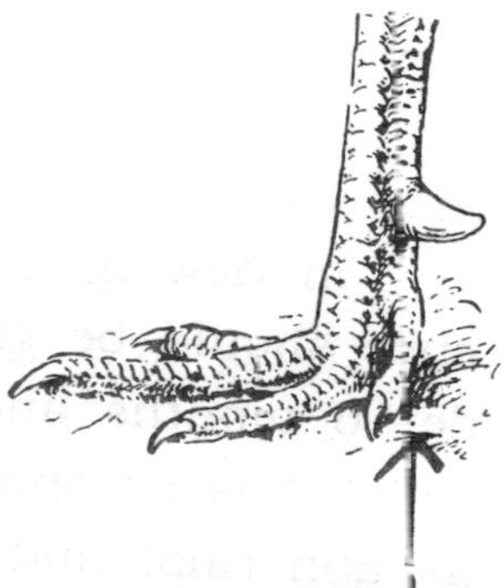

Condition Known as Duck-Footed
This is a very serious fault in OEG

Figure 4.9 The Foot of the OEG Bantam

Spur: Low on the leg.

A fine spur should appear on each leg of the cock bird and if it grows too long should be removed, which can be done quite easily by taking off the outer shell of the spur using a pair of pliers. Those with no experience should consult some-one with the necessary knowledge. Sometimes the hens are spurred, although not as frequently in bantams as in large OEG.

The *standards* are silent on spurred large, hens, but the Game enthusiasts of old preferred these because they were regarded as a sure sign of ferocity and 'Gameness'. On this basis spurred hens in bantams should not be penalised because, as far as possible, bantams should resemble the large OEG.

In viewing the spurs of a male bird, allowance should be given between young birds under 6 months and fully mature cocks. A **cock** which has no spurs would certainly be regarded as abnormal and therefore not be worthy of a prize. A late hatched cockerel may be late in growing his spurs so be sure to allow for this fact when viewing a young bird.

Carriage: Bold, smart, the movement quick and graceful, proud and sprightly as if ready for any emergency.

The saucy, combative antics of the bantam cock are recognized in the English language. Without the aggressive, show-off behaviour the character is lost. The reference to *being ready for any emergency* applies to the days of cockfighting when the unready cock was quickly disposed of by his opponent. These days the 'carriage' description means **style,** exhibiting the showiness which is an essential part of the OEG bantam.

In the Oxford *standards* around 10 points would be allowed for carriage; in the Carlisle standard a total of 40 points is given for shape and carriage so it will be seen that shape counts for around 30 points. In the OEG Bantam Club the body, breast, back and belly count for 20 points.

Without the essential **style,** an OEG bird is lacking in the character which is an inherent mark of its make-up and therefore, despite other qualities, should not be regarded as an ideal bird. Far too many birds lack this vital character and thereby risk the omission from the prizes. Unfortunately, too many 'all-round' judges see the heart-shaped body and nothing else and award on that single characteristic alone.

Handling: Flesh firm, but corky with plenty of muscle. This to be considered an important factor.

The judging of OEG must be done visually and by handling; a bird is held in both hands and 'balanced' in the hands to see whether it is light and corky. Bantams may appear heavy for their size and too fat and obviously is a condition to be avoided by giving plenty of exercise out of doors.

Plumage: Hard, glossy and firm.

The feathers of OEG should be hard and fine without excessive fluff at the base. They should fit close to the body forming a sheaf which is tight fitting. Any faults in the feathers should be considered when making awards. Broken feathers, mis-shapen, irregular growth, missing feathers and any other irregularities should be penalized. Dirty feathers and the presence of mite should also be set against a bird. The OEG bantam should not be as hard feathered as a Malay or other Asian Game bird.

In bantams the **correct colour** is an important consideration (see next chapter) and any departure from the colour *standard* should be penalized.

Weight: Cocks; it is considered not desirable to breed cocks over 6.5lb. (The reference is to **large** Game)

For bantams about 25% of the weight of the large breed is taken, which means in the region of 20oz for females and 25oz for cocks. It cannot be exact because large Game do vary a great deal from around 4lb up to 7lb or even more. The *Poultry Club Standards* now give the weights in metric measures as **Male** 620 to 740 g and **Female** 510 to 610g, but we have given the Imperial measures to allow comparisons with other countries.

Some argue that bantams are getting too large; many of the 'Old School' prefer the smaller, compact bantam rather than the thick set bird which shows signs of Indian Game blood. Certainly any movement towards larger, coarser birds should be resisted, but neither should there be an attempt made to produce smaller and smaller specimens because this generally weakens the constitution.

What is the Correct Size?

One fancier likes the very heavy-set bird because it usually has a bulky body, including a broad and prominent breast. Another likes the diminutive bird which is very small and compact. It is to this latter group of fanciers that the word 'cobby' is very pertinent; this means small and compact, displaying a roundness in shape, yet proportionally having all the requirements for the ideal shape.

In assessing the shape and the weight it is essential to avoid coarseness in body, head and legs. An OEG bantam should be refined and elegant and yet comply with the *standard.*

The cock's weight at around 26oz or just under provides a guide to the correct size, but balance and harmony should be present, along with style if a bird is to be regarded as top class.

The difficulty is how does the judge react when there is a small or large bird being shown. Rarely do judges weigh birds, but go on 'impression'. There seems no reason why this should not be done. OEG are taken out of the show pen anyway so a steward could quickly check the weight. If more than 20% away from the standard weight then it would seem reasonable to penalize the bird in question.

***The ASP penalizes for under- or over-weight at a stated rate; ie, 1 point for each 2 oz under and half a point for each 1 oz over.**

The Points System

The award of points for the main features is long established, but if it is to be used fully there is need for a score card which is used by judges. In fact, not many judges appear to favour this scheme and, although they are guided by the scale, tend to favour an 'all round appraisal' rather than a detailed analysis. In any event, it will be appreciated that the judging of birds is very much a subjective appraisal and the break-down into elements can only be used to give the comprehensive summary **(the prize giving)** at the end.

If a **Judging Card,** is used, **with standard and actual points compared,** this allows comparisons to be made before awards are given. It brings an **element of objectivity** into the whole process of judging which is after all a comparative analysis - deciding whether one bantam or another comes nearest to the *standard* being used.

Any competent judge knows he must analyse the subject in detail to be able to state with certainty how near each bird comes to the **ideal**. Often one bird will have an ideal colour, but has a poor tail, whereas another has a splendid head, but has rather coarse legs. Only by considering the points stipulated in the *standard* will it be posssible to consider the relative merits and reach a rational decision.

The Scale of Points

The 'Scale of Points' which follows is an integral part of the Carlisle Club *Standards* and should be used in conjunction with the written descriptions. Other scales vary in detail, but all agree that **shape** is the most important. In the Old English Game Bantam Club, which is adapted from the Oxford Club Standards, a different break-down is given. Number 1, Shape and Carriage 40 points, are shown as Body, Breast, Back & Belly 20 points and Carriage & Action 9 points. Item 2 are identical at 15 points. However, for item 5, Colour and Plumage, 20 points, the OEGBC gives only 9 points.

SCALE OF POINTS

1. Shape & Carriage	40
2. Handling & Condition	15
3. Head & Eye	10
4. Legs & Feet	15
5. Colour & Plumage	20
	100

These variations simply give a different emphasis to different features, but overall they are the same.

HEALTH & CONDITION

Bantams which are not in top condition should never be shown because it is unfair to the bird and may spread infection to other exhibits. In addition, there is little hope of winning with any bird which mopes in the show cage and show none of the alertness and vitality which are essential features of OEG bantams.

A rule that should always be followed is that any bantam *which is obviously suffering from a serious complaint* should be disqualified. This does not apply to birds which are not absolutely fit on the day; that is, they are simply not up to peak condition, but on a better day might well have taken a top award. Such exhibits would be placed in order of merit after reducing the points for ***Handling & Condition,*** shown as 15 maximum, above.

Winner
B & T Wilkinson
Dark by modern requirements.

Splendid Hen
George Harding

Supreme Champion Carlisle

Pullet

Both J & M Stalker - shows the trend towards lighter coloured birds.

Figure 4.10 Wheaten Females

Generate argument on correct colour for hackle, tail and wings. Tendency is for very light cream coloured bird with few dark feathers (if any) to win.

5

STANDARDS II: COLOURS

SHAPE BEFORE ALL?

There is a belief in many circles, both Game and non-Game fanciers, that shape comes first and nothing else matters. Many have been known to scoff and assert that shape wins irrespective of the mongrel appearance of a bird as depicted by the colour. Thus it has been stated that: *....there are now (unfortunately) a large number of bantams of bastard colouring winning prizes in the show pen, simply because they are of good shape and feather. Most of these ought, properly, to be rejects and are derived from 'off-coloured' breeding pens.**

Not all would agree with the conclusion reached, but caution must be exercised or all purity of blood and breeding will be lost at the alter of the worship of 'shape'. Being able to breed the standard colours, which show that

---

*** *Bantams for Everyone*, H Easom Smith, BPH. A general guide to bantams. It must be stated that the author was not a Game fancier and believed that OEG received too much attention and favour, but the book is excellent.**

breeding pens are being carefully selected, is one of the prime aims of setting the *standards*, thus ensuring that the pure varieties are kept intact. If 'off-colours' are to be included in the **Any Other Colour** or **Any Other Variety** then, provided there is no mixing with the standard colour classes, no harm is done. However, all things being equal, when selecting the top awards, it seems proper that purity of colour should take precedence over a bird which has been produced by some chance breeding on the basis of shape alone.

The Standard Colours

The recognized colours are referred to as the standard colours, but unfortunately, these differ from one country to another; thus in the USA* we have a **Ginger** OEG bantam which rarely appears in any other country. In the UK there is the **Furness** which is not in the USA. Even the colours are not always given the same description so a pure comparison is not simple.

The Standard colours are as follows:

USA Standard Colours

According to Fred P Jeffrey (ibid) , Secretary, American Bantam Association, there are

*** American Standard of Perfection (ASP in text); the American Poultry Association,inc.**

29 recognized colours in OEG bantams. The ***American Standard of Perfection*** is not so explicit. This lists 19 as follows:

Birchen
Black
Black Breasted Red
Blue
Blue Breasted Red
Blue Golden Duckwing
Blue Silver Duckwing
Brown Red
Crele
Cuckoo
Ginger Red
Golden Duckwing
Lemon Blue
Red Pyle
Self Blue
Silver Duckwing
Spangled
Wheaten
White

The **ABA** list includes, **additionally:**

Barred
Black Tailed Buff
Black Tailed Red
Black Tailed White
Blue Wheaten
Buff
Self Blue
Silver Blue
Silver Wheaten
Splash Wheaten

British Standard Colours

These are derived from a number of sources and tend to follow a similar pattern to the USA but with more emphasis on the natural colours; eg, Black-Reds. Variations in a main colour, such as Black-Red, may be regarded as sub-divisions of that main colour, and therefore not entitled to be classified as a recognizable, standard colour. Thus we can have a Blue-Tailed Wheaten, but this is simply the result of the introduction of Blue colour into the Wheaten.

A list of the main colours found in Britain is now given, but do not expect to see these at all the shows; indeed, some may not be shown except at the large national events. They are as follows:

Black
Black-Reds (Partridge and Wheaten)
Blue
Blue Duns
Blue Reds
Brassy Backed
Brown Reds
Crele
Cuckoo
Duckwing - Golden
Duckwing - Silver
Furness (or Furnace)
Grey
Pile
Polecat
Spangle
Splashed
White

A number of these will not be listed in the standards as such, but will be found to exist from examination of the show schedules of the major, national shows.

Old English Game Club (Carlisle)

The standard colours are now described with notes as previously for the main standards.

Points of Colour in Spangle Game Cock

Face: Bright red.

Eyes: Red, both to be alike.

Neck Hackle and Saddle: Dark red finely tipped with white.

Breast & Thighs: Black , finely and evenly tipped with white.

Back & Shoulders: Dark red, finely tipped with white.

Wings: Wing bow dark red, finely tipped with white, with a rich dark blue bar across, finely tipped with white; secondaries deep bay intermixed with white, bay predominating; primaries black intermixed with white.

Tail: Sickles and side hangers black tipped with white; straight feathers black, intermixed with white.

Legs: White or yellow.

This is a beautiful little bird which has been popular from the early days of showing. The dark red is a rich burgundy, although the depth of shade varies. The 'spangles' are small tips at the end of the feathers. The ASP defines these as **diamond-shaped** on the neck hackle and wings, and **V-shaped** on the back; in each case there is a thin black line dividing the Spangle from the remainder of the main colour. The number of spangles and their distribution varies and if too frequent the birds are too "gay" and resemble a Belgian Barbu d'Uccle Millefleur bantam, which very attractive in itself, is not complying with the standard because chicks from such a cock would produce light coloured hens. In fact, by accident I produced such a colour from a chance mating by a wheaten hen and Spangle cock.

The legs and beak should usually be white, but occasionally come yellow (British) and the beak may be a horn colour.

Generally, the off colour Spangles, such as Black or Dark Brown should be discouraged, although it is sometimes necessary to use these crosses to fix a particular point and then the alien colour is bred out. If a bird has white blotches or other large spots of white, usually irregular, this colour is known as 'Splashed' and should not be allowed in the Spangle classes. Often such birds come from attempting to breed Blues, when Black and White Splashes result.

Colour - Spangle Hen

Face: Bright red.

Eyes: Red both to be alike.

Neck Hackle: Golden red, Streaked with black, finely tipped with white.

Breast and Thighs: Dark salmon colour, finely and evenly tipped with white.

Back and Shoulders: Dark partridge coloured feathers, finely and evenly tipped with white.

Wings: Secondaries, dark partridge intermixed with white, partridge predominating; primaries dark, intermixed with white. All other feathers dark partridge colour, finely and evenly tipped with white.

Tail: Black with partridge coverts, finely and evenly tipped with white.

Legs: White or Yellow.

A nice even partridge colour is required; some specimens are a reddy brown, almost grouse coloured. If too light the spangles do not show up so well as the darker colour.

Faults: Male and Female - Hackle very light or extremely dark; lack of spangles; excessive white in wings; dark coloured legs and beak; black lines below spangles too broad so they are really black spots;eye wrong colour - yellow, white or very dark.

Points of Colour in Black-Breasted Red Cock (Partridge)

Face: Bright red.
Eyes: Red, both to be alike.
Neck Hackle and Saddle: Orange red, free from dark feathers.
Breast and Thighs: Black.
Back and Shoulders: Deep red.
Wings: Wing bow, deep red with a rich dark blue bar across; secondaries, bay colour on outer web; primaries black.
Tail: Sound black with lustrous green gloss.
Legs: White, yellow or willow.

This is the natural colour of a Game bird, based on the red Jungle Fowl. The description follows the traditional pattern with the breast being stated first followed by the remaing colour; thus Black-Red should be read as *Black in breast and red on hackle and shoulders.*

The colour is a delight to behold being a very good, deep colour, darker than the Wheaten Black-Reds However, it is not as dark as the Black-Breasted Black-Red or Dark Red, seen in large Game, where the hens are a purple Black. Sadly, this colour is very rarely, if ever, seen in bantams.

Colour - Hen Partridge

Face and Eyes: Same as cock.
Neck: Golden red and streaked with black.
Breast and Thighs: Shaded salmon colour.
Back and Wings: Partridge colour, to be as free from rust and shaftiness as possible.
Tail: Black with partridge coverts.
Legs: White, yellow or willow.

This is a lovely colour and more pleasing in my view than the wheaten which is now so pale as to be simply a creamy white. I won first prize at the Alexander Palace National show with a lovely hen which was a natural partridge colour, very evenly marked and no trace of red in the wings which is a common and serious fault caused by breeding with cocks that are too red in the back and saddle. "Partridge" is a combination of black and medium brown, finely blended and stippled into a partridge colour. Only by observing the colour is it possible to appreciate its beauty and subtle blending.

Faults: Cock - Very dark red in cock; brown feathers in cock in breast or on thighs; white in wings.

Faults: Hen - Ruddy colour in wings or across back; white in wings, signs of wrong colour in

Figure 5.1 Partridge Hen
(1st Prize National Show - bred by author)

hackle such as grey stripes, indicating a cross with duckwings.

Faults: Both - Any deviations from standard colour in feathers or leg or eye colours.

NOTES ON BLACK-RED (WHEATENS)

Males: This is really a lighter version of the normal Black red. Originally the Wheaten colour was a pale fawn and the cocks were somewhat patchy in colour, depending on the female line. There is also a colour known as 'Clay' which is a yellowy fawn, which, again, the *light* Red cock will produce. These days in bantams there must be absolute purity of colour or the hens will be unacceptable as wheatens. Accordingly, unless double mating is practised, one pen for hens and the other for cocks, the cock's colour must be a very light shade of orangy-red colour.

The ASP stipulates a **light orange** for the hackle and saddle and this colour is usually in accord with the colour now prevailing with light wheaten hens.

Females: The overall colour should be a creamy wheaten and modern winners display a very pale shade with very little in the way of darker markings. The "golden red" of the hackle has become a slightly darker wheaten, which is inevitable with the dilution in colour in the body of the hen.

Points of Colour in Bright Red Cock (Wheaten)*

Face: Bright red.
Eyes: Red, both to be alike.
Neck Hackle and Saddle: Light golden red, free from streaks.
Breast and Thighs: Black.
Back and Shoulders: Bright red.
Wings: Wing bow, bright red, in other respects similar to Black Red.
Tail: Similar to Black Red.
Legs: White.

Colour - Hen Wheaten

Face and Eyes: Same as cock.
Neck: Golden red, free from striping.
Breast and Thighs: Light Wheaten.
Back and Wings: Wheaten, level on colour.
Tail: Black with a shading of wheaten corresponding with body colour.
Legs: White.

Hen: Clay

Similar to Wheaten only darker or harder in colour.

Faults: Male Wheaten - Dark feathers on head or in hackle; colour other than black on breast or thighs; legs other than a pinky-white.

Faults: Female Wheaten - Dark colour or shading on body or hackle; wrong colour eye.

Also known as** ***Black Breasted Light Red.

Points of Colour in Brown Red Game Cock *

Face: Gypsy or red.
Eyes: Dark, both to be alike.
Neck Hackle and Saddle: Lemon or orange streaked with black.
Breast and Thighs: Black, laced with brown.
Back: Lemon or orange.
Wings: Shoulders and wing bow lemon or orange, rest of wing black.
Tail: Black.
Legs: Willow or black.

This is not a popular colour in bantams and they have tended to be on the large side. Nevertheless, they are worthy of a following because they often have a wonderful, lustrous black feather throughout the body and possess great vigour. The wings are 'crow winged' meaning there is no separate colour at the bottom of the wing (the bay).

A great deal of latitude is allowed in this colour which is natural for a bird which has lacing ranging from lemon to a deep mahogony brown.

Faults: Plumage which is not full of lustre, mismarkings with other colours such as white in wings; pale eye and light coloured legs.

*** Originally the colour was a brown breast, which later became a brown-laced breast and now a black breast is accepted.**

Colour - Hen Brown Red

Face: Gypsy or red.

Eyes: Dark, both to be alike.

Neck Hackle: Lemon or orange, striped with black.

Breast and Thighs: Black laced with brown.

Body: Black.

Tail: Black.

Legs: Willow or black.

The overall colour is black or a black and dark-brown, mottled and merging together, to give the impression of black. The striping and hackles in the cock should be a similar shade, although when showing this will not always be apparent.

The lemon colour is very distinctive, but the golden brown is more in keeping with the natural colour of the Brown Red.

The plumage should be harder than the Black Reds, showing a gleaming purple and green tinge. The ASP equates the *standard* for Brown Reds with Modern Game colours, but this is too rigid for the British *standard.*

The gypsy face is a dark colour, sometimes described as 'mulberry' which is not strictly correct because the colour should be a dusky brown and not purple.

Faults: As for males. Also colour of body which shows Partridge markings.

Points of Colour in Blue Red Cock

Face: Bright red.

Eyes: Red, both to be alike.

Neck Hackle and Saddle: Orange or golden red.

Breast and Thighs: Blue, medium shade.

Back and Shoulders: Deep or bright red.

Wings: Wing bow, deep or bright red with a rich dark blue bar across; secondaries, bay colour on outer web; primaries blue.

Tail: Blue.

Legs: Any self colour.

Colour - Blue Red Hen

Face: Bright red.

Eyes: Red, both to be alike.

Neck: Golden red, streaked with black.

Breast and Thighs: Shaded salmon colour.

Back and Wings: Partridge colour, intermixed with blue.

Tail: To correspond with body colour.

Legs: Any self colour.

In effect, these are Black Reds which have been crossed with blues to replace the Black with Blue. They are beautiful bantams and have a strong following. The comments made on Black Reds (Partridge) apply, but substituting Blue for Black. There is a latitude allowed with the leg colour because of the Blue influence, but white is preferred.

Colour - Blue-Tailed Wheaten Hen

Face: Bright red.
Eyes: Red, both to be alike.
Plumage: Similar in all respects to Wheatens with the exception of wing primaries and tail shaded with blue.
Legs: White.

This is a Wheaten colour, but there is this permeating shade of blue, especially noticeable in the tail. It is very attractive, but should be kept as a separate colour for the Blue does spoil the more conventional Black Red/Wheaten colour if mixed.

Colour - Self-Blue Cock & Hen

Face: Bright red.
Eyes: Red, both to be alike.
Plumage: Blue, medium shade.
Legs: Any self colour.

Blues should be the one, single colour, an even slate grey. They vary because, being bred from a Black base the exact shade depends very much upon the intensity of the Black in the mixture. In fact, to maintain any Blue colour it will be necessary to have a cross with Black and then select the Blues required.

Colour - Pile Cock

Face: Bright red.

Eyes: Red, both to be alike.

Neck Hackle and Saddle: Orange or chestnut red.

Breast and Thighs: white.

Back and Shoulders: Deep red.

Wings: Wing bow, red with a white across; secondaries, bay colour on outer web; primaries white.

Tail: White.

Legs: White or yellow.

Pile Game, probably derived from 'Pied' because of the mixture of colours, are very attractive birds. They are rather like the Black Reds, but instead of the Black colour there is White. The hackle and shoulders show off best in a deep scarlet - known in large Game as 'Blood-Wing Piles', but in reality there are many different shades varying from a yellow to the deep-red type. An occasional cross with a Black Red Partridge keeps the colour strong.

In size they tend to be diminutive, a sign of being well established as a bantam. Their draw back is the light plumage which will certainly call for regular washing for shows.

White legs tend to be preferred although the yellow-legged varieties are very attractive.

Colour - Pile Hen

Face: Bright red.
Eyes: Red, both to be alike.
Neck: Lemon.
Breast: Salmon colour lighter towards thighs.
Back and Wings: White.
Tail: White.
Legs: White or yellow.

Pile hens appear white, but in reality are a creamy-white colour, being influenced by the yellow which is in the colour scheme. The salmon colour of the breast varies depending upon the intensity of the Cock's colours. In a Custard Pile (yellow hackle) the female will tend to have a pale fawny breast.

Some latitude should be allowed in the colour, but if too pale in the breast and hackle the hen appears as a badly coloured white and therefore should be penalized. The purity of colour is important and therefore too much of a departure should preclude a first prize when better coloured, standard-type birds are present. Where the colour is altogether wrong the hen should be regarded as being in the wrong class.

Note: In the ASP (American Standards) the colour is known as a 'Red Pyle' , thus giving a different spelling.

Colour: Golden Duckwing Cock

Face: Bright red.

Eyes: Red, both to be alike.

Neck Hackle: Creamy white.

Saddle: Orange or rich yellow.

Breast and Thighs: Black

Back and Shoulders: Orange or rich yellow.

Wings: Wing bow, orange or rich yellow; wing-bars black; secondaries, white on the outer web; primaries black.

Tail: Black.

Legs: Any self colour.

Duckwings are a very attractive colour, but not the most popular in bantams. The distinctive, bold yellow combined with the black and the steel-blue of the wing bar (resembling a mallard duck's wing bar) give a very pleasing combination. They breed quite true, but any cross with Black Reds to deepen the yellow tends to giving the hens a rusty colour across the wings.

NOTE: The distinction between Golden and Silver is one of degree. The cocks are a yellow colour in Goldens with hens inclined to have a darker tinge to the overall body colour. Side by side the hens of the two colours can be distinguished, but, separately, it is sometimes difficult.

The Silver Duckwing cock is quite distinct

Colour - Silver Duckwing Cock

Face: Bright red.
Eyes: Red, both to be alike.
Neck Hackle and Saddle: Silver white free from dark streaks.
Breast and Thighs: Black
Back and Shoulders: Silver white.
Wings: Wing bow, silver white; wingbars black; secondaries, white on the outer web; primaries black.
Tail: Black.
Legs: Any self colour.

Colour - Duckwing Hen

Face: Bright red.
Eyes: Red, both to be alike.
Neck: Silver, striped with black.
Breast and thighs: Salmon colour.
Back and Wings: Steel grey free from rust and shaftiness.
Tail: Black. Coverts corresponding with body colour.
Legs: Any self colour.

and his silvery colour should not be yellowing in any way. As indicated, his matching hen should be a distinctive silver-grey. The fawn tinge in the Yellow Duckwing hen should not be present in the Silver hen. The **overall colour** is not a self-colour, but made up of a mixture which give an overall impression of grey (Cf. Partridge).

Colour - Crele Cock

Face: Bright red.
Eyes: Red, both to be alike.
Neck Hackle and Saddle: Chequered orange red.
Back and Shoulders: Deep chequered orange.
Wings: Wing bow, deep chequered orange, with a dark grey bar across; secondaries, bay colour on the outer web; primaries dark grey.
Tail: Dark grey.
Legs: White preferred.

The Crele is similar to the Cuckoo, rather like a Barred Plymouth Rock. As the Poultry Club point out the Crele is a 'chequered' bird other than the barred grey. Theoretically, there could be any colour and Red or Golden are as above, and silver also exists. However, it seems the Red type is the most acceptable for a Crele and the Grey/Blue for the Cuckoo or Mackerel.

Not surprisingly it is related to the Black Red and a cross with a cock of this colour with a Crele hen usually produces Creles.

The aim is to achieve a good strong colour, but ***double mating*** may be necessary for producing both male and female of top quality. With a normal breeding pen the pullets bred will tend to be rather weak in colour.

Colour - Crele Hen

Face: Bright red.
Eyes: Red, both to be alike.
Neck: Lemon, chequered with grey.
Breast and thighs: Chequered salmon colour.
Back and Wings: Chequered blue-grey.
Tail: To correspond with body colour.
Legs: White preferred.

Colour - White Game
Cock & Hen

Face: Bright red.
Eyes: Red, both to be alike.
Plumage: All over pure white.
Legs: White

As would be expected, both sexes have to be the same white colour. In reality, though, there is a difference because the cock is harder in the feather and his hackle tends to be slightly off-white because glossy feathers have a different hue from the softer feathers of the female.

White are beautiful, but difficult. They tend to be softer in the feather than the darker birds and the frequent washing for shows make them softer still. Nevertheless, for the patient they are worth the effort.

Colour - Black Game
Cock & Hen

Face: Bright red.
Eyes: Red, both to be alike.
Plumage: All over glossy black.
Legs: Any self colour.

Usually the Blacks are Blue/Black bred so they are reasonably soft in the feather. Occasionally they are seen in a Melanistic Black when they are hard and glossy in the feather.

Any feathers with colours other than black will call for a penalty, possibly disqualification if in a Black only class. Really top quality birds are to be found in this colour; small, cobby and tame.

Any ticking or lacing or other marks should lead to disqualification if being shown in a self-colour class.

Figure 5.2 Examples of Black OEG Bantams

Cock - Top - rather large tail for today's winners.
Bottom - Hen & Cock showing fullness of Breast without deep keel.

6

STANDARDS III: COLOURS

IMPORTANCE OF COLOUR

The fact that colour is very important has been noted. When a specific colour is listed in a show schedule, then any birds ***entered in the wrong classes should be disqualified.*** There are ***standards*** to follow to show the fancier what the colour is meant to represent.

These may be found in the appropriate *standards* issued by the various clubs throughout the world. Where colours emerge which are not listed, then it is usual to show these in mixed colour classes. Only when numbers are adequate to make a class worthwhile will it be possible to have a separate class for the more unusual colours. Sometimes even the popular classes have to be merged, so Black-Red Partridge and Black-Red Wheatens are shown together, but this is not a sound idea because a judge tends to prefer one or other and, subconsciously, a decision has already been made before the judging starts.

RECAP - Colours

Colours recognized in Britain, the USA and other countries are as follows (see pages to 78):

Black*
Black-Reds (Partridge and Wheaten)*
Blue*
Blue Duns
Blue Reds* (also Blue-tailed Wheaten)*
Brassy Backed
Brown Reds*
Crele*
Cuckoo*
Duckwing - Golden*
Duckwing - Silver*
Furness (or Furnace)
Grey
Pile (Pyle)*
Polecat
Spangle*
Splashed
White*

Off-colours (mainly USA) are as follows:

Birchen
Blue Golden Duckwing
Blue Silver Duckwing
Ginger Red

---

* **Those marked with an asterisk have a detailed *standard* in the preceding chapter.**

Barred
Black Tailed Buff
Black Tailed Red
Black Tailed White
Blue Wheaten (see Blue-tailed Wheaten)*
Buff
Silver Blue
Silver Wheaten
Splash Wheaten

OTHER COLOURS

Colours not listed in the **Carlisle *Standards*** call for comment:

1. Blue Duns

This is a colour which causes much confusion. Strictly 'Dun' means a yellowish or greyish brown colour, but nowadays in Game Fowl it has come to mean a Blue fowl with other colour splashes or markings although, theoretically, we could have Silver Duns or Honey Duns which should mean silver/dun and honey/dun respectively. Generally a reference to *Dun* means a Blue Red.

2. Furness and Brassy Backed

The Furness is a bird with a single colour overall, such as black or blue, but the shoulders and back exhibit a dark flame colour, but in that part only. In the normally coloured cock the hackle is the same colour as the shoulders, but with a Furnace this is not the case because the hackle is the same as the rest of the body.

The *hens* for the Furnace cock are almost self-coloured, but have a slight mixing of other colours and often shafts show in the feathers. The breast may be a fawnish colour mingled with the darker colour.

In effect the Furnace is a self colour bird with a *diluted form of colour gene* which makes the red in a Black Red. In its weak state it covers the shoulders only.

The Brassy-Backed colour is like the Furnace, but the shoulders are a golden-yellow (Brassy).

3. Polecat

The reference here may be to the colour of the ferret known as a Polecat which is a mixture of yellowish fawn and black. It is a mixture of colours, rather similar to the Furnace without the red saddle.

The female tends to be a yellow-ochre colour with black or brown markings and shafts a darker colour.

4. Grey and Birchen

Greys are a popular colour in large Game, but not with bantams. They are as for Brown reds, but instead of the yellow or orange the colour for hackle and shoulders is a clear silvery-grey with a black stripe. A laced breast is desirable, but following the large Game is not insisted upon.

Birchen is the name given to Modern Game of a Grey colour, but it is a misnomer because 'birchen' is really a brown colour. The two (Greys and Birchens) are virtually the same. The ASP gives the hackle and saddle as 'white with narrow, dark stripes', but in reality the colour is more of a silvery colour, as would be expected with a bird which is predominantly Black with dark eyes.

5. Ginger Red (Gingers)

This is a beautiful colour which has been in existence

in large Game for many generations. A painting by Herbert Atkinson shows these magnificent birds in all their glory. The overall colour is predominantly gingery-red. The colour in the cock is like the Black Red with a ginger colour instead of black. The hen is an overall ginger colour (mixed in colour like the Partridge hen) with lacing on the breast. The colour should be very bright and glowing.

The ASP depicts a different shade to the British birds theirs being more subdued in colour.

They are related to Black Reds and Brown Reds and will cross with the latter to produce darker coloured birds. The original Brown Reds of long ago had brown/mahogany breasts and possibly it is from this source that the colour came.

6. Blue Golden and Silver Duckwing

Instead of Black for normal Duckwings substitute Blue.

7. Black Tailed Buff, Red and White

These appear to be self colours except for the tail. They are recognized by the American Bantam Association.

8. Silver Blue

This is similar to the Grey given earlier, but instead of the predominantly Black colour there is Blue. The hackle and saddle are silver in the cock and so is the hackle of the hen.

Where there is a metallic wing bar and coloured wing bow they become Silver-Blue Duckwings.

9. Other Self Colours

Where birds are single colour they are referred to by that colour; eg, Buffs. If the cock has more than one colour then it cannot be called a 'Self Colour' because the cock's plumage determines the colour.

10. Other Colours

Colours like Black Breasted Black Reds (Dark Reds) exist in Large, but are very rarely seen in bantams.

COLOURS OF EYES, LEGS & BEAK

The general rule with Game fowl is that the eyes, legs (shanks and toes), beak and even toe nails should match. This rule is no longer followed rigidly, especially with bantams. The following guide lines should be useful:

1. Preferred colour for eyes for all medium or light coloured bantams is red. Leg colour and toe nails should be white.

2. Dark coloured birds should have dark eyes and dark legs (black, grey, carp or willow) to match the overall colour; for example, Brown Reds and Greys. However, this does not always occur because certain colours have white legs; eg, Blues and Blacks, although a really dark Black (not Blue-bred) could have dark eyes and legs (slate to match body). Sometimes in Blacks the scales show dark markings and these are quite natural, although a pure colour is preferable.

3. Yellow legs are permitted on some colours such as Piles, but white tends to be preferred. In Duckwings some variation is permitted, but white preferred.

4. A yellow, pearl or daw eye should be regarded with suspicion because this is a certain sign of an alien cross. The bold red eye always looks so much better except on the dark birds as mentioned for Brown Reds and Greys.

Faults in Colours

The colour should be as specified in the *standard* and non-compliance should mean a penalty. In the **Scale of Points,** Colour and Plumage count for 20 points (Carlisle). However, this does vary and may be as low as 9 points, which is very much under-rating its importance. The fact is that any bird in a particular class should be within a reasonable distance of the specification.

Faults which should be penalized as as follows:

1. Self Colours

Any colour which is different from the self colour.

2. Specific Colour Variations

If a cock is a Black Red then the black should be free from all other colours; eg, lacing or brown or white feathers.

White in wing feathers of any bird except white birds, but varieties such as Piles, Spangles and Light Reds may have white wing feathers; in such birds white is a predominant colour; these should not be excessive in number (see Spangles note 4. below).

2. Markings

Spangles, speckles or other markings which exist, when these are *not* a feature of the main colour; eg, Duckwings with spangles on breast, when, as a result, it is clear that the birds in question have had Spangles in their ancestors.

4. Spangles

These should be spread in a regular fashion. In the cocks,there should be a double row of spangles on each wing bar and even spangling on each secondary feather. The primary feathers should not be too white, the ideal being a width of about two inches (5cm). More than that and the wing and spangling is too white.

Many Spangles used to suffer from being short of spangles and the cocks being too dark which has to be watched very carefully.

Figure 6.1 Spangle Bantam Cock (Show Champion - Sam Lean) Note the extremely short back and double row of spangles on wing bar.

Faults

Very few spangles; cock much too dark.
On shape not cobby enough; far too 'angular'.

Figure 6.2 Spangle which is too Dark

Figure 6.3 Golden Duckwing
Full fronted and compact (R Douglas)

Figure 6.4 Shapely Black Red Cock
A winner for M Crawford & Son

7

GETTING THE RIGHT SHAPE

SHAPE IS ESSENTIAL FOR OEG

Different fowl have different outlines of the body and feather formations which make up the characteristic of the particular fowl. Accordingly, **contrary** to what is sometimes suggested, that only Game fanciers '**go for shape and shape alone**', it must be appreciated that other breeds must also exhibit the correct shape or they will not comply with the breed ***standard***.

Possibly the difference is one of degree, but even then, in some *standards* , the 'body, breast, back and belly' are given 20 points (British Poultry Club/Oxford *standard*). As noted earlier, the Old English Game Club gives 40 points for Shape and Carriage. The ASP* states in general terms that shape and type in any breed must constitute two-thirds of the total points to be awarded.

Differences between ***standards*** are often due to a breakdown of characteristics in different ways rather than a wide divergence in opin-

***Standards of the American Poultry Association.**

ion. Accordingly, it is necessary to consider the detail of separate standards before jumping to conclusions.

The Essential OEG Shape

The essential cone shape has been described in the earlier chapters dealing with the characteristics and ***standard*** for OEG. It is now necessary to consider the following:

1. Overall shape or silhouette, which is the essential profile of the breed for each sex.

2. The breeding of the essential shape and how it can be improved.

The silhouette can be seen from Figure 7.1 and 7.2. In each case the impression of breed is seen at a glance. Notice the smallness of the tail in the British birds and the way this grows from the body to give a streamlined effect. It is this attention to the rear of the bird which has resulted in the very narrow structure of the bird. Obviously, this is why the tails have become smaller and more refined with weak sickles.

With the American and Australian birds, which follow the large type of Game, the tail structure is wider and stronger and the result is a greater width at the tail end. The feathers are wider and a full pair of sickles is present.

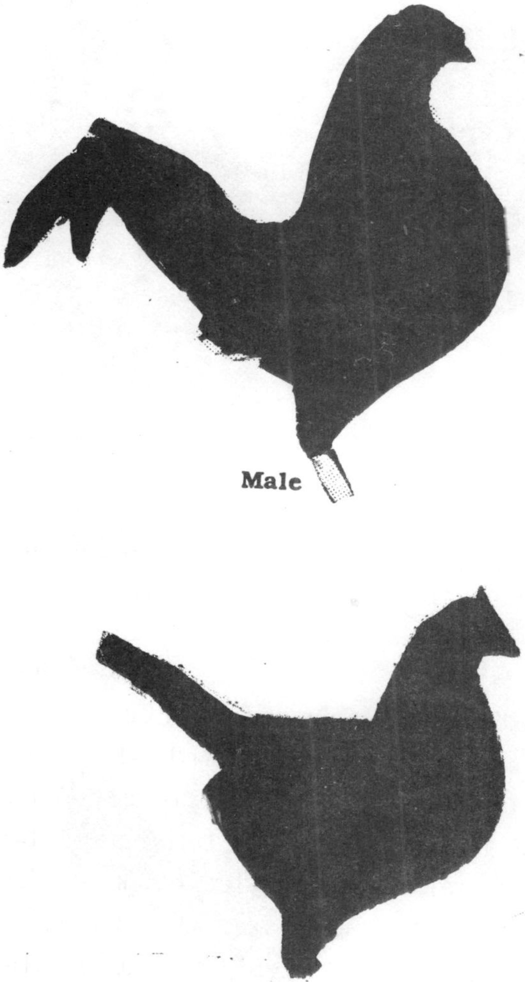

Figure 7.1 The Silhouette of British OEG Bantam Cock and Hen

Male
Dwarf-like with rather large tail; not diminutive like British counterpart

Female
Model of large Game hen, but smaller

Figure 7.2 Shapes for OEG bantams (USA)

Where Does the Shape Come From?

The bone structure and the way the legs are set and, accordingly, the stance of the body, are the determining factors, all from the genes in the birds put together for breeding. Without the correct combination of genes there can be no standard-type birds; only those which possess the desired combination will produce the necessary shape.

The shape is not simply the body; it is that but much more. The way it moves (carriage) and the various appendages such as head, neck, legs and so on all affect the appearance of the bird. Features which should receive attention are as follows:

1. **Shape of body outline.**
2. **Balance or symmetry and *shortness of back.***
3. **Size.**
4. **Head.**
5. **Beak.**
6. **Comb and related appendages such as wattles and ear lobes.**

All these are covered in the Chapter, Standards I, which shows the requirements for each feature. It is the harmony of all these which is vital to the appearance of OEG. Getting shape is relatively simple, but we do not want stodginess or clumsiness, both of which are contrary to the essential character of OEG. A thickset bird has shape, but does not usually have grace of movement or the vital spark which is so much a part of OEG bantams.

The fundamental rule therefore is to breed from those parents which comply with the standards and do not have any major defects which will certainly show in the offspring. A sagging abdomen, showing between the legs, is an abomination which spoils the essential upward curve of the body so birds with this fault should not be allowed in the breeding pen. Old hens in lay will tend to show this tendency because laying develops the abdomen. However, this can be allowed for and hens with excessive 'under carriage' should be avoided. It may also be a sign of over-weight and sometimes it is excessive feathering which exaggerates the bulge.

In Game birds it is always stated that hens have the most influence on shape. This has been my experience; therefore make sure the hens are exceptionally shaped.

In addition, there are various principles which should be followed:

1. Learn the *Standard* for OEG bantams and understand what it means.

2. Only breed from those birds which comply with the *standard*.

3. Mate a cock with one or two hens, no more or control will be difficult. Remember that the cock is one side of the breeding equation so make sure (if possible) that his mother was a winner or a potential winner.

4. Endeavour to spot the correct combination for breeding winners. It is usually suggested that any weakness on one side should be compensated on the other; thus a broad breasted hen with deep keel should be mated to a cock with a shallow keel, thus arriving at progeny with the correct depth of body.

Whilst there is sense in this suggestion unfortunately it does not always work. Some breeders prefer to match parents with as few faults as possible thus avoiding the possibility that the fault will be carried into the next generation.

5. Select on a very strict basis on criteria based on the standard and the type of bird which is winning prizes. The idea that there is a static, never changing *standard* is an illusion because the interpretation relies on the judges. A reader must be ready to move with the times and be up front with the prize winners.

6. Only by developing a stud which has all the necessary qualities will consistent winning be possible. This may take four or five generations, or even longer, to establish so that similar birds are being produced each time.

A recognizable strain may take many years to establish so do not be deceived by those who allege they have developed a new, winning strain in a season.

I knew one famous breeder who was so confident that he would sell his old stock at the end of the breeding season and rely on stock just bred. This is not to be recommended because some parent stock should always be kept as a foundation stock for corrective breeding should the season's breeding go wrong. Some breeders have access to related stock, thus being able to get back to the original bloodline.

SHORTNESS OF BACK

A feature that is discussed and commented on when viewing birds is '**shortness of back**'; ie, the body viewed from the top (the desired heart shape), measured from the base of the hackle to the base of the tail. This feature is difficult to achieve and yet keep the overall harmony of a bird. If very short, the bases of the hackle and tail just about meet, and this produces a bantam with very exaggerated shoulders and narrow near the tail. Many judges criticize such a short-backed bird as lacking balance and this is justifiable. Nevertheless, the hunt continues for the extremely short back.

Recent winners in hens have tended to appear to have longer backs because of the angle of the tail. Instead of the 45° angle it is in the region of 30°, or even less and this appears

*** The method of measuring the angle of the tail is covered earlier.**

to make the back longer. On the other hand, if the tail is above the 90° angle (squirrel tailed) this will appear to make the back shorter, but is unacceptable and calls for disqualification.

The aim should be to achieve a tail in the region of the ideal of 45° because this gives a positive marker for the end of the back. Also, if the tail is at too low an angle, the type is approaching the style of many Asian breeds which are required to have a drooping tail.

IN-BREEDING OR LINE BREEDING?

Inbreeding and **line breeding** are closely related. Both involve the use of related birds for breeding. Thus the mating of brother and sister or male parent back to daughter is usual.

The effect is to reduce the number of ancestors and to maximize the **effects** of genes possessed by both sides. This means that desirable features (*standard* features) can be emphasized and thereby gain points at shows. Two birds bred from related stock which have small tails will emphasize this feature more the next time round; other features can be brought out in the same way.

Line breeding follows the same pattern, but involves the selection of an individual cock or hen and perpetuating his or her genes by continuous use of that bird in breeding programmes.

Once established the line can go on, even after the original male is lost. This is done by breeding the best daughters to the male for a few generations and then selecting a new male, which by this time will have a considerable percentage of the make up of the original male. Alternatively, the line might be established on the female side, thus acknowledging the importance of the hen in the shape.

The main problem with inbreeding is the fact that it brings out the recessive genes as well as the dominant and therefore undesirable features may appear.

Selection of strong stock with no sign of deterioration is vital. I inbred large Blue Game for 20 years without difficulty, but after that time breeding became very difficult; the vigour had been lost and no other self-blues could be found to introduce new blood. Fortunately, with bantams, related blood can be found so the stock can be given new blood when the old becomes weakened. It is important to keep to **related**, outside stock or the entirely new blood may produce many new problems from the appearance of recessive features, including very unexpected colours.

If new blood has to be brought in, then expect to have to stabilize the effects over two or three generations. After that the new blood may have given much more vigour so the strain can continue. In this connection some breeders believe that the introduction of a female may be less damaging than a male.

Once the desired characteristics have been stabilized the pen which produces the best results should be continued as long as possible. This follows the principle of 'progeny testing' whereby the results of breeding are recorded and the outcome recorded and studied.

If too many faults emerge from a breeding pen then some variation must be tried by introducing a new cock or hen.

Use of Different Colours

Many Game fanciers introduce new blood by breeding with a different colour. Thus in Pile Game new blood might be a Black-Red Partridge cock. This would introduce fresh vigour and deepen the colour, but expect strange colours to appear until the genes have 'settled down' again.

There is nothing more irritating to the dedicated Game breeder, contrary to the adage that a Game bird can be any colour, to see a multitude of mixed, random colours emerging from the breeding pens. Spangles on black, brown or even wheaten birds; blue-tailed colours of all varieties, and Brown Red or Greys

BREEDING RECORD

Pen No..50.......... **Year:...19......**

Cock:Desc.:.BlkRd........RingNo.V12...............

Age:.2...

Females:......2.......... **Age:2....RingNo...V41/46..**

EGGS SET Date:.3/3/19--......No..12............

Hatched:...10.........

Details of Chicks:

Ring No.	Size	Type	Colour	Comments
X22	Lge.	Cobby	B/R	Rather Dark Male
X23	OK	Cobby	Partge	Wings Ruddy ?
X24	OK	Tall	B/R	Good; may fill out.

Note: all chicks hatched would be recorded as above

Culled/Sold(RingNos.)..............................

Age Culled:

Comments...

Figure 7.3 Breeding Record

This may be combined with an incubation record or done separately which allows more detail

with white wings or mismarked feathers of all descriptions. It may be a way to improve shape or add new blood, but it is also a sign that the breeder is a dabbler who crosses this way and that without thought to purity of colour.

A much better plan is to stick to the accepted colours, but where new blood is essential to bring in colours which do not clash. Thus Brown Reds and Greys are related; Black-Red Partridge are related to Piles and Golden Duckwings, and Blacks, Blues, Black and White Splashes and Furnaces are variations of the same basic colour. Others could be mentioned, but this list should indicate that certain colour do harmonize whereas others are difficult to get rid of, once introduced. Spangles should not be crossed with other colours and neither should Blues for this colour seems to linger on for ages and the blue tint is for ever re-appearing.

BREEDING METHODS

With OEG breeding it is very usual to have a cock or cockerel and one or two hens; in this way it is easy to see the source of the eggs and obviously the best birds can be selected. The old breeders would see whether a pair 'nicked'; ie, whether this was the correct combination to produce the winners. Once established the combination would be continued as long as possible.

A virile cock can run with more hens than two or three as suggested, and when not breeding seriously this is the practice followed having about six hens. Sometimes a cock may be allowed to run in alternate pens every other day or for a few days at a time. This is the 'Stud Method', allowing separate pens for, say, two hens in each, with the cock in a cockerel shed in the middle with a trap door opening (pop hole) on either side to let him out to the allotted pen.

The time for breeding is usually from February to the end of July, chicks born in the March to May period get the best of the weather and grow very quickly so they are strong and fit. The earlier chicks might catch the Autumn shows so there is an incentive for the earlier hatching; with a 25 watt bulb giving a few hours extra daylight, which also gives more feeding time, it is surprizing how quickly chicks come on. However, there must be warm, dry quarters inside because attempting to rear chicks in cold, inclement weather is quite impossible and is unfair to the mother and chicks. If artificial rearing is followed then a cosy compartment in a well insulated shed will suffice.* Late hatching is not usually worthwhile.

*** Space does not permit full coverage of incubation and rearing and readers are advised to see *Artificial Incubation & Rearing* and *Natural Incubation & Rearing* by J Batty.**

Is Double Mating Necessary?

The practice of **double mating** for showing is necessary when a single breeding pen will not produce male and female winners so separate pens are essential. Usually the reason for its adoption is because the ***standard*** details a specific colour or other characteristic for, say, the male bird and in breeding this the female is given an undesirable feature. In other words, the ***standard*** is not realistic and does not allow both sexes to breed 'true' from the same pen.

We can see the principle of 'feature clashing' with OEG when a brightly coloured Black Red cock is used to produce chicks. Quite often the Partridge females will have a reddish-brown colour across the wings and this will persist because of the extra red in the genes of the male. Such females are not likely to be show winners because they do not have a good, even partridge colour. In such cases, double mating could be practised, but generally it can be avoided by a more careful selection of the male bird; the brightness of the red is a matter of degree and cocks can still win with slightly subdued colours.

My experience suggests that double mating should be unnecessary for OEG bantams and a breeder who introduces it will not get better results than those who breed along normal lines.

INTRODUCTION TO GENETICS*

There is no need to be a scientist to breed prize-winning bantams. However, for those who wish to understand the subject there are many excellent books which can be studied.

Modern genetics started with an Austrian monk named Gregor Johann Mendel, who died in 1884 and who discovered, through his gardening experiments, in which he grew sweet peas, that - **like produced like** - tall peas produced tall offspring and dwarf plants produced dwarf plants. When he crossed the two varieties the result was ***not*** an intermediate size, but tall plants.

Subsequently he crossed the second generation and obtained both tall and dwarf in the ratio of 3 tall to 1 dwarf. The first discovery from this was that some features are produced more than others (**dominant** characters), whereas the remainder are **recessive** characters.

With all living things we start with a cell which is made up of a **chains of genes**, known as chromosomes. These occur in pairs and there are probably 23 pairs, with one pair determin-

*** This introduction is a simplified approach to the genetics of poultry. Readers should consult a specialized work for an authoritative coverage.**

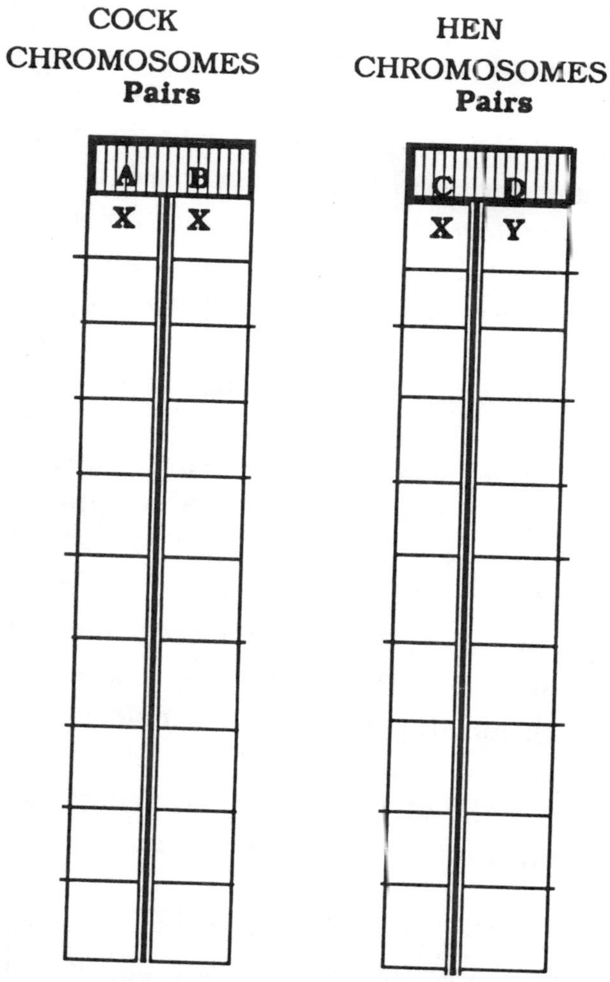

Figure 7.4 Chromosomes

The blank squares represent the different genes. The actual number will be normal (autosomes) whereas the remaining pair will be the sex chromosomes X and Y. It should be noted that a male fowl will have not less than 17 autosomes, plus one pair of sex chromosomes.

ing the sex. In the case of a bird it is the female which determines the sex.

When the sperm of the male meets that of the female the chromosomes mix together so that characteristics from each parent appear. This mixing together produces variations in the structure of the chicks so that shape, colour, feather texture, eye colour and so on will be affected.

Certain genes are **dominant** so when these occur this is the character which will appear. The **recessive** characters remain hidden, but may appear at a later time with a subsequent mating.

The sperm of the cock joins the ova of the hen and two new pairs are formed, thus determining the sex. The remaining chromosomes also marry up and part of the female joins part of the male to form cells which determine the character of the offspring. The precise result will depend on which factor is dominant over the other.

In the first mating, as indicated, the dominant characters will emerge, but in the second and subsequent matings recessive characters will also appear. When dealing with colour, say, Black and White, since neither colour is dominant, a mixture of colours will be bred, including Greys (Blues), Blacks, and Black and White Splash. When Blue is crossed with Blue we can expect 50% Blue and 25% Black, the remainder being White (or white and black splashes).

In effect, this illustrates what occurs when the chromosomes meet and re-form. Part of the male and part of the female will bring out the different combination. However, with some colours there may be genes present which affect the normal expectation patterns. In fact, the Red Pile may be a recessive White although they can produce Black Reds.

Mendel's principles, although modified from practical findings, still apply today. It is possible to state the following 'laws':

1. *Law of Dominance*

The dominant factor appears in the first breeding and then the recessive appears in the ratio of 1 to 3.

2. *Law of Incomplete Dominance*

This is where the genes are both dominant so a mixture appears as noted above for breeding Blues.

3. *Law of Independent Assortment*

Many of the characters segregate and they regroup in proportion to the original numbers and this usually applies where hybrids are being bred.

Many experiments have been carried out to test the theories and to confirm that the expectations do appear. The great difficulty is that we can never be sure which genes are present in specific birds, especially the recessive genes.

The male and female have '**gametes**' (reproductive cells) and when the two are joined in the egg this is known as the **zygote** which means a fertile egg. It is from the zygote that the combined chromosomes emerge.

Because there are hundreds of characters in a bantam, it follows that the total number of pairs of chromosomes are responsible for a number of each. The genes give rise to the many variations which emerge from the breeding.

Although the first law (above) purports to suggest that 'like produces like', in reality this is not the case. A fundamental principle with poultry breeding has been the recognition that, although matings will produce similar birds, they are never exactly the same. **It is on the constant variation which is present that the numerous breeds have been developed.** The breeder has to try to recognize the characters that are present (aided by suitable records) and then use this information to develop a winning strain which can then be improved upon. Only then can the breeder be satisfied.

8

MANAGEMENT

NATURE OF MANAGEMENT

OEG bantams are quite hardy and will thrive under any normal conditions, whether in an enclosed run or living free range in an orchard or field. The main danger comes from predators such as the fox and therefore it is important to keep a watch for any dangers; above all see they are locked indoors each night.

The various matters requiring attention under ***management*** are as follows:

1. **Accommodation of various types**
2. **Feeding**
3. **Breeding**

These are covered in that order.

ACCOMMODATION

With OEG bantams it is necessary to think in terms of winter and summer housing. In summer they can have an outside run or have access to the paddock or yard, preferably where they can scratch for food and eat green food such as grass and chickweed. No special protection is required, but a roosting shed which is dry and free from damp is essential.

When winter comes the bantams may still be let out in the elements, but they are better in a covered run with leaves, peat moss, or wood shavings in which they can scratch. Moreover, although not the best of layers in winter, in the right surroundings they will still lay and lead a normal active life. This is much better than having to cope with rain, sleet and snow which they do not really like.

Essentials in the house or run are the following:

1. Perches which are not too smooth so the birds can grip easily and fixed above the ground so they can fly up. Do not have them too thick or their toes will be spoilt by trying to grip. If possible, have a stepping effect by having a series of perches. Generally they will keep flying up until on the highest perch.

2. Food hopper and drinking fountain. The birds must have a supply of fresh water and once each week give the container a thorough clean. The hopper can contain pellets for ad lib feeding (see *Feeding* below).

3. Nest boxes so that the hens lay inside, lined with hay, straw or shavings.

Figure 8.1 Typical Small Sheds for Bantams
A run may adjoin each house which will be suitable for a trio of bantams or more depending on overall space.

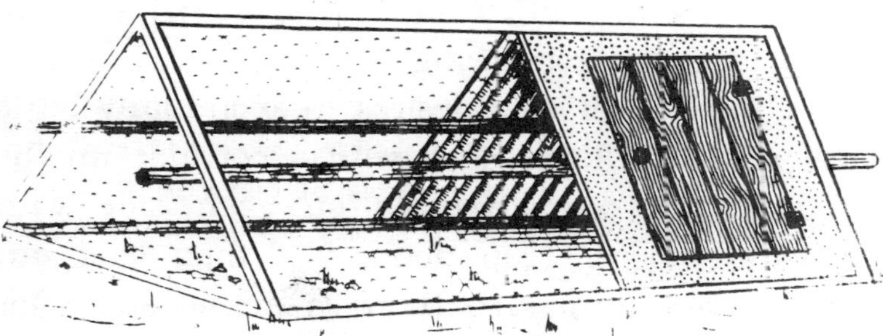

Figure 8.2 Fold Unit
This can be moved regularly to avoid sour ground and may be used for adults or for raising growing stock

Using A Large Shed

For those with limited space or for winter quarters a garden shed is ideal. A shed 8 ft. X 10 ft. can be divided into two compartments with a door between and two trios can be kept, one in each section. However, be sure to have hardboard along the bottom of the partition because bantam cocks like 'to have a go' and therefore they are better separated. Also include a platform shelf about 2 ft. wide across the back and about 5 ft from the ground. This acts as a droppings board and a nest box can also be included. The height gives the birds exercise as they fly up to reach the food and perches.

The floor should ideally be concrete or very strong wood (tarred) to keep out rats and mice. This should then be covered with garden soil and leaves and other materials, added to form deep litter for scratching.

The fancier who desires to keep many birds will require a large shed with a corridor up the middle and compartments on each side with separate doors. Trap doors will then lead out into separate runs so there is ample space for breeding pens. In addition, if showing, a Penning Room or shed will be essential for training and getting birds into condition.

FEEDING

'Keep them hungry, but feed them well' should be the principle for OEG bantams. This means they are given adequate food according to the season and whether the hens are laying or not. The easy way is to have a hopper full of poultry mash (dry powder) or pellets; this is known as ***ad lib feeding*** and birds can thrive quite well on it. However, it is more suited to laying stock than exhibition bantams, especially the hard-feathered type. What is needed is variety and presented in such a way that the birds are eager to eat and they are prepared to scratch and exercise not simply sit on a perch all day and eat to capacity morning and evening.

A menu can take many forms, but it must be practical. Pellets of the small type for layers can be given in the morning by scattering in the litter. For those who are away early a hopper can be used for birds to help themselves, but the ration for the day should be watched. Feed mixed corn in the evening and if the birds eat a good handful each you will know they are not eating too many pellets. If they are sluggish and leave the corn then only a limited quantity should be left in the hopper.

Some fanciers advocate a handful of corn

or two in the morning and moistened mash in the evening, but this is a matter of preference on the part of the person involved.

Water is vital because without it, even for a day the birds will suffer. If for any reason birds are deprived of water - usually due to a water fountain leaking - the combs of the birds will turn a smutty colour and the birds will not lay. In fact, if left short of water, hens may resort to eating their own eggs, a habit which is very hard to break once established.

Grit, both soluble (limestone and oyster shell) and insoluble (small flints) should be available. The former is for calcium and the latter is for grinding up the food in the gizzard. Both aspects are essential.

Greenstuff such as lawn clippings, leaves, chickweed, dandelions, and other wild, edible weeds should be given on a regular basis. If the birds are running outside they will pick their own.

At moulting time give the birds extra protein to help with feathering, but remember OEG bantams are supposed to have tight glossy feathers so too much protein might spoil the result required. On the other hand, a complete diet of wheat only, advocated by some fanciers, whilst giving very hard feathering leads to brittle feathers which may not last out the show season. Try to give a balanced diet which will give a long lasting effect and keep the birds in good condition.

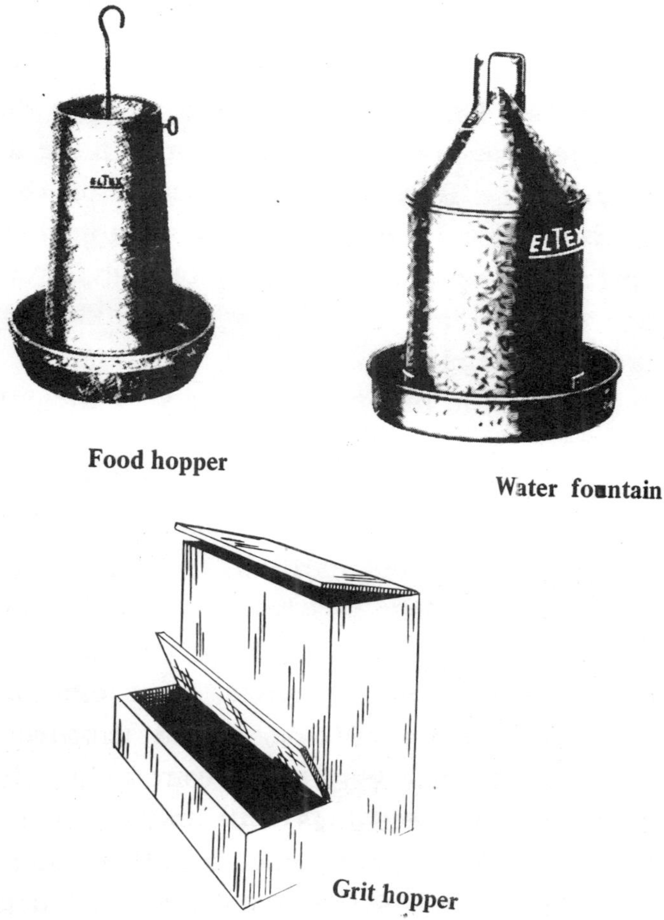

Figure 8.3 Food, Water and Grit Hoppers

BREEDING

Breeding is the most enjoyable part of bantam keeping. It presents a challenge and there is always the problem arising which makes the whole process so interesting.

In Chapter 7 the approach to selection of stock for producing winners is considered. The reader is advised to understand what is the **aim** or **objective** and then attempt to achieve it. This may be the **Best Black Red cock** or **the improvement of a colour** or **the production of a better shape and shorter back**; whatever it is, try to keep this in mind when making up the breeding pens.

Start the plannning process in the Autumn and put the pair or trio of birds together about November when they have moulted and are getting into condition. Select them carefully along the lines suggested earlier.

Once in their separate pens make them as comfortable as possible and introduce more layer's pellets and a variety of titbits; a little bread soaked in milk seems to encourage the appetite. If early eggs are desired then have a light on in the evenings which will stimulate the laying and give more time for activity and feeding. Make sure all the birds are in good condition and for early fertility use a young cock. The older bird may not be fertile until later in the season.

Yearling hens (no older early in the season) should be used and use the older, tried hens in another pen later on. However, do not introduce new hens into an established breeding pen because fighting will occur between hens. If any changes are to be made then select an entirely new house and run and start off from scratch with the new breeding trio.

During the breeding season keep the birds active and feed layer's pellets of the small size; let them on to a grass run if possible because this really gets them into top condition. However, if the weather is frosty, or wet and very inclement the birds are better kept inside where they will thrive quite well on deep litter. There should be good light from a window and adequate fresh air. Removable shutters are a sound idea for these allow the worst conditions to be kept out and can be removed when the weather gets better.

The bantams can be allowed to sit, but a better idea is to use a special broody which is small, yet will cover a dozen eggs. A Silkie or Silkie cross is ideal; a pure Silkie is excellent provided her soft feathers at the front are removed because chicks get trapped in them.

A small incubator is an alternative, but is not really worthwhile unless at least 30 eggs are to be set at regular intervals. Remember the chicks have to be reared so hatching them alone is only part of the problem. Early in the season an incubator may be the only method available, but remember it only becomes a viable proposition if the procedures are properly organized.*

The essentials for incubation and rearing are as follows:

1.Collect eggs daily and store them in a cool place in trays or boxes containing shavings. If dirty when collected wash very carefully in water to which has been added a mild disinfectant to sterilize the shell.

2. Mark the eggs with a code for the pen/birds and the date. If to be incubated put an X and Y on opposite sides of the egg to aid with the turning process; it shows which eggs have been turned, which should be done twice daily to prevent the yolk sticking to the membrane.

3. Pre-heat the incubator and make sure it is warmed up before loading with eggs. These should be no more than 7 days old. The temperature will be in the region of 102 deg. F with the thermometer just above an egg.

*** Space does not permit coverage so readers wishing to incubate and rear should refer to *Artificial Incubation & Rearing* by the author.**

4. The eggs should be tested (candled) at 7 and 14 days. Any that are clear or addled should be discarded.
5. Expect the eggs to hatch at 19 to 20 days and from the 17th day do not turn or interfere in any way. The moisture is usually increased just before this time.
6. Once hatched move the chicks to a brooder or broody hen which has been sitting for 10 days or more. Place the chicks under her and gradually move the eggs under her into the incubator. Watch carefully to make sure she takes to them (*move quickly if she pecks them*), but do nothing more for 2 days except to feed the broody. If a brooder is used (possibly infra-red lamp) pre-heat to 95 deg. F. and then drop by 5 degrees each week by raising the lamp.
7. Feed chick crumbs initially and at 14 days introduce broken corn and chickweed. After about 6 weeks introduce grower's pellets and continue with the corn. Once fully feathered the chicks can be put in an outside run and they will grow very quickly. Make sure the ground is fresh to avoid diseases. For water use a chick-size water fount and change the water every day.

OEG bantams are small and delicate for the first week, but after that thrive with no difficulty. They simply require sound accommodation and food and water. Do not let them perch too early or the breast bone may be affected.

Special Points To Watch

When breeding OEG bantams there are one or two matters to watch:

1. If a broody hen is to be used try to avoid hens which are large and clumsy or those with feathered legs. On the latter a cross Silkie is better than a pure bred because of the fluffy feathers and the feathered legs which may cause problems such as scaly legs (caused by a tiny mite) which can be passed to the chicks. If it does occur treatment with sulphur ointment is advisable or paint on paraffin (make sure this does not touch the skin because it burns).

2. Chicks that are mopy and appear to be having problems with growing wing feathers (they show very unevenly) are probably suffering from coccidiosis which is a disease of the intestines. Fortunately, the standard chick crumbs contain a *coccidiostat* which will prevent too many losses. If there are a number of cases a drug should be introduced into the water (obtainable from a farm shop or by prescription from your vet.)

3. If one of the birds in the breeding pen becomes ill then isolate and try to determine the problem. Feeding on soft food may be the answer, but do not attempt to breed from the breeding pen until the matter is cleared up.

4. If eggs are clear (ie, not fertile) it may be the cock which is too old or too young. Early in the season a cock may be out of condition, but will improve later. Sometimes old

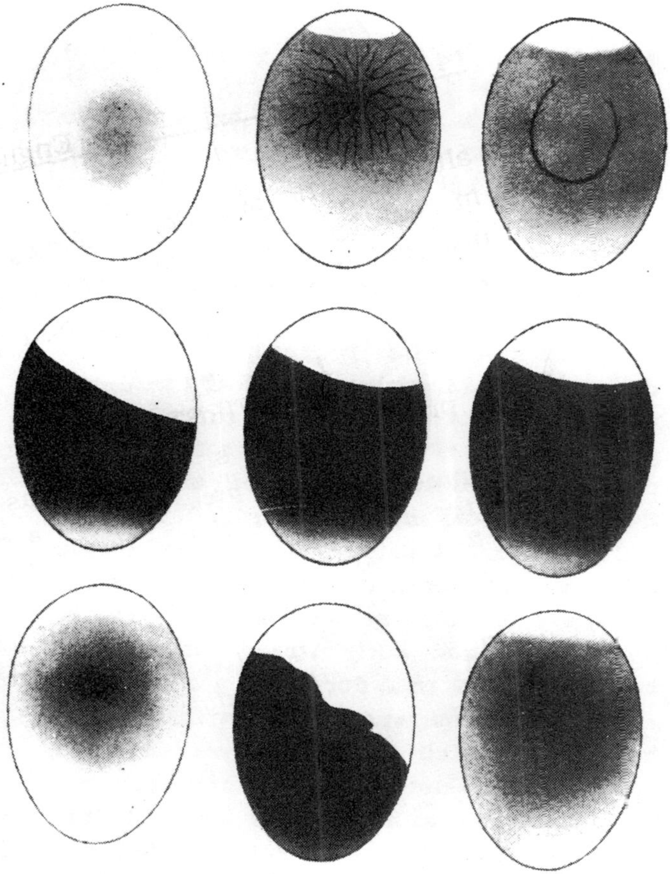

The egg during incubation: *Top row* (7th day) left, clear egg; centre, fertile egg; right, broken yolk. *Middle row* (14th day) left, egg dried too much – too little moisture; centre, correct drying; right, too little drying – excessive moisture. *Bottom row* left, dead embryo about 5th day; centre, ready to hatch 20th day, right, embryo died on about 14th day.

Figure 8.4 Eggs at stages of Incubation

Test the eggs with a bright light at 7 and 14 days. The air space and growth of dark area must be observed carefully

hens are too fat or bully a cockerel so the eggs fail to be fertile.

5. Do not be in a mad rush to breed early in the season. Often early chicks catch the autumn shows, but those hatched a few weeks later may progress just as fast; better weather, especially sunshine can make a tremendous difference to fertility and growth of chicks.

6. Watch stock birds to ensure there is no mite on them. Examine around the vent and remove any dirty feathers and rub with sulphur ointment. Provide a dust bath so birds can dust themselves in dry soil or soil and fine ash. A box under cover, filled with the soil is all that is needed. If feathers become affected with feather mite (feathers eaten away) try bathing the bird in a solution of warm water and fairly strong disinfectant, but make sure this does not get into the eyes. In addition, spray with an insecticide, but, again, watch the bird's eyes and also ensure they do not breathe in the fine spray.

7. At about 3 months of age it will be better for males and females to be separated because they thrive better that way and it avoids the fights and squabbles that may take place.

8. Bantams can be reared intensively, but there should be direct light, other than through glass, which implies mild weather conditions. Cod Liver Oil added to the food (a few drops per day and vitamins – Abidec) can help the process.

9

THE SHOW SCENE

The Display Window

Shows have been popular since Victorian times. They provide a 'show window' for the fancier at which he can compete with his best birds and see other breeders' stock. Viewing other birds indicates the progress being made so that the achievement of the *standard* is shown as a possibility. The opinions of experts, including those of the judge, may also be obtained on an informal basis, although the judge must never be embarassed by argument or by threats; he or she is under no obligation to give reasons for decisions made.

COMPLYING WITH THE STANDARDS

As shown earlier, ***standards*** exist for physical properties and for colour. Birds shown must comply with the appropriate *standard*, although obviously, **absolute perfection** will never be achieved. In the case of OEG bantams, although the ***standard*** follows the large Game, in reality there are differences, especially in the overall carriage and type, and particularly the tail.

FIRST STEPS

The procedures for showing are as follows:

1. Regard show preparation as a *continuous process* and not simply one which starts a few days before the show date.

2. Learn the *standard* so that all its definitions, colours and other features are understood. Also appreciate how the judges are interpreting the *standard* at the present time.

3. Get the birds tame so they can be handled. Examine and handle them regularly. In this connection have:

(a) A Penning Room or shed in which birds can be conditioned and trained which contains:

(b) Individual show-type cages where the birds can be kept for training and when getting ready to go to a show.

(c) A judging stick to train the birds to stand correctly to enhance the presentation.

(d) Medicines such as dusting powder, ointments, vaseline and dubbing scissors.

(e) A sink if possible, although this is not vital because bowls of hot water can be carried into the shed. Washing in the kitchen might be feasible, but can dis-

turb household harmony.

4. If possible be a member of the Old English Game Club and the Old English Game Bantam Club, as well as any local clubs which run shows.

5. Obtain show schedules in time for entry of the best birds. Do not take any bantams to the shows which are out of condition, especially if they are mopy. Those with broken and/or broken feathers should not be shown. Cocks *not* dubbed cannot be shown.

6. For any important shows birds should be washed and thoroughly dried before taking to the show. This usually means washing them about 7 days before the event. Dark coloured birds may escape a wash except for legs and face, but light colours cannot be missed.

7. On return from the show there are differences of opinion on procedures. Some fanciers take the view they need exrecise and should be put back in the pen with the birds they left behind. Others believe that rest in a show cage with soft food is better. OEG bantams are very hardy so which ever method is adopted the return to normal life as quickly as possible, with plenty of exercise, is vital.

Some of these aspects are now considered further in the sections which follow.

PLANNING THE SHOWS

If **regular showing** is to be a feature of being a fancier it is necessary to plan ahead. A listing of the shows to be attended is advisable together with dates and when entered. (see Figure 9.1)

The steps or stages involved and the selection of a team of birds will also be an essential requirement or there is a danger that your time will be overloaded or you may miss some of the important events.

Some new fanciers believe that they must take as many birds as possible without regard to condition or show potential. **This is quite wrong.** It is necessary to examine each bird against the ***standard*** and if it falls short in a number of aspects it should not be shown. In fact, if very poor, unless there are special circumstances, such as the existence of a valuable blood line, not in other birds, the birds with many faults should be **culled**; ie, removed from the stud by selling or other method - some child may like a pet. However, be careful not to dispose of pairs or trios of the winning strain (eccept at the correct price) because they may produce winners in the future for the new owner.

Show Planning Schedule

Main Shows	*Dates*	*Entry*	*No. Entered*
OEG Club			
OEG Bantam			
National			
Federation			
Rare Breeds			
3-Counties			
LOCAL SHOWS			

NOTES:

Figure 9.1 Show Planner

Selection List

Show:......................... *Date:...........*

BIRDS ENTERED

Bird/Ring	*Class*	*Preparation Req'd.*
Pile Cock\ 38	8	Wash; Feed linseed in Moult.
Blk\Rd Part (F)	12	Wash legs; Dub immediate.

Notes: Dubbing to be done at least 4 weeks before show.

All birds to be shown must be through moult or find substitute.

Review 2 weeks before show and make final selection before washing.

Figure 9.2 Selection List for each Show

WASHING FOR SHOW

Birds which are white or light coloured should be washed. Darker coloured birds may also be washed, but this is not essential for every show. The **legs** should be made spotless and usually these are washed on a regular basis throughout the year.

The **face and beak** should be wiped over with a soft cloth and a wipe with a little fine oil is advisable. However, do not use any cream or preparation which is likely to cause a reaction. Even women's cleansing cream can start the surface of the comb flaking, which does not look good at a show. The intention is to have the face and comb red and bright so a wipe over when the birds are being penned might be the best way.

The start of washing consists of a thorough cleansing of the legs, where necessary scrubbing with a nail brush, using a washing-up liquid and then rinsing with cold water. Do not attempt to show birds with ***scaly legs*** or with some form of disfigurement.

Next comes the **main wash** which involves taking each bantam and immersing it thoroughly in fairly hot water to remove all the dirt. Washing up liquid or a mild shampoo should be added to make it 'soapy', and then the bird is washed. The wings should be spread out on a draining board and scrubbed with a small scrubbing brush so all dirt and stains are removed. The remainder is rubbed with the fingers until all traces of dirt are removed. During this process keep the bird in the water up to the head, but not allowing the water to touch the beak or eyes.

Once washed quite well the bird is transferred to warm, clean water, and all traces of detergent or soap are rinsed away. If necessary a **further bowl** is then used to give a final rinse, again in warm water.

The bird can then be dried with a soft towel. A hair-dryer may be used to complete the drying or some form of drying cage or basket may be employed. Some have special drying boxes rather like the traditional hospital cages which are heated by an inbuilt tubular heater or element. Keep in a warm room until the drying is complete.

Points to watch are as follows:

1. Do not use harsh detergent or shampoos which will take all the oil out of the feathers and make them quite 'soft'.

2. Allow a number of days for the drying to be completed. A bird not properly dried cannot show itself to advantage and will not feel confident. Wash feet and legs once more

immediately before taking to the show because they tend to get soiled. Dirty feet and legs give a poor impression when the judge handles the bird to assess its merits. Load into the show hampers just before leaving.

Figure 9.3 Dubbing of OEG Bantam Cockerel

DUBBING

Dubbing is the process of removing the wattles and comb from a **male** bird. It is carried out at around 6 months of age, when the comb is fully developed and the 'operation' is done by using a pair of small, curved surgical scissors obtainable from a chemists. In some countries a vet. must perform the operation. However, it is a **very simple** procedure which has been done decades and the bird feels no pain or ill effects. **What is important is for the person doing the dubbing to understand the procedures and to be quite confident.**

Dubbing can be a work of art when the bird is left with a **half-moon portion** of the comb and a smooth neck without any ugly bumps or blemishes. A true Game fancier takes pride in well dubbed birds. The comb should not be cut very low to the base in the fashion of Modern Game; neither should the comb be cut across without any attempt to provide shape. These unorthodox cuts often leave a cock looking 'sneaky', whereas he should look bold and defiant.

Procedure

The cockerel is held in the left hand close to the body. Switching the hold to the body by holding the bird very close the head is held with the thumb and top two fingers. Then with the right hand proceed to cut off the **wattles** so the neck is perfectly smooth. This is followed by **starting at the back of the comb and then cutting to the front** to curve down to the beak to form the half moon shape. It is over very quickly - just about two minutes. The procedures need to be mastered so it is essential to be shown and, if necessary for two people to be involved.

For showing, OEG bantam cocks must be dubbed. This is an ancient tradition and once done requires no more attention. Moreover, should they be involved in accidental fighting little or no damage will be done.

TRAVELLING HAMPERS

Sending or taking valuable birds in cardboard boxes is fine provided they are correct for the job. Once used they can be burnt so no disease can be carried. Those who show regulary will find show hampers more suitable. They are compact, are easily handled and transported, and are well ventilated.

Make sure the **shavings** are removed from the hamper when the birds are returned from the show and put in new wood shavings immediately before the new show.

Some fanciers give the birds water in the hamper, but unless there is a special watering device the water may spill out. A container of moistened bread, attached to the side, is an alternative method. **It is essential to give water, especially on a long journey in hot weather.**

Figure 9.4 Show Hampers

Cardboard boxes of different shapes and sizes are difficult to transport to shows; a special hamper is much better.

Figure 9.5 A Penning Room

This was the penning room of a well known fancier; it is vital that birds are show pen trained.

SHOW SCHEDULE & REGULATIONS

Fanciers should obtain the show schedule for each show and send it off in good time. If for a 2-day show, well away from home, it will also be essential to arrange accommodation to stay overnight. If there are any queries on entries contact the Show Secretary quickly and make sure you are making the entries correctly. Also ensure that you comply with any regulations on hygiene or supplying with food and water (usually birds in a show are fed and watered by the stewards). If there are regulations on such matters as vaccination make sure this is done in good time.

A RECAP OR REVISION EXERCISE

Throughout the book the many features required for the **ideal** bird are examined. When the day of the show has arrived the judge will decide whether your bird deserves a prize and he will do that by comparing your entry with the others in the class.

He or she will follow a routine of deciding such matters as the following:

1. Is the bird in the correct class.

2. Does it comply with the requirements of the *standard* and does it have any major faults (listed below)

Black Reserve Champion
M Evans & Son

Spangle Cockerel
W E Carpenter

Black Pullet
W Crellin

Grey Hen
Uncommon colour
B R Skinner

Figure 9.6 A Variety of Winners

3. Is the bird healthy.
4. When handled, is it reasonably tame and is the breast even (no indentations), does it feel firm, but not fat or over-weight.
5. If it is a male has it been dubbed correctly.
6. Has there been any faking or malpractices on the bird*.

Faults Which Will Be Penalized or Disqualify

1. Duck-footed*.
2. Stork legged, stiff legged or distorted in any way from normal stance*.
3. Deep keeled or not straight*.
4. Weak Headed (narrow and sneaky).
5. Foul Feathering or badly feathered in specific colour class*.
6. Lacking Shape.
7. Lethargic and lacking sprightliness of OEG.
8. Faulty Carriage.
9. Narrow at shoulders.
10. Wide at tail with large tail or no tail*.
11. Any disfiguring features such as scaly legs or sores or bumps. Thick instep and square shanks (OEG should be round)*.
12. Lice, mite or the existence of any other pests.
13. Rounded or Roached Back*.
14. Much Overweight*.

15. Faulty Wings; goose winged, low carried wings, split wings, feathers missing, wrong colour in primaries; eg, in Spangles too dark or too much white*.

Any feature which does not comply with the ***standard*** should call for a penalty and those asterisked above (*) could require disqualification.